# Nanostructure Based Sensors for Gas Sensing

# Nanostructure Based Sensors for Gas Sensing: From Devices to Systems

Special Issue Editors

**Nicola Donato**
**Sabrina Grassini**

MDPI • Basel • Beijing • Wuhan • Barcelona • Belgrade

**MDPI**

*Special Issue Editors*

Nicola Donato
Department of Engineering
University of Messina
Italy

Sabrina Grassini
Department of Applied Science and Technology
Politecnico di Torino
Italy

*Editorial Office*
MDPI
St. Alban-Anlage 66
4052 Basel, Switzerland

This is a reprint of articles from the Special Issue published online in the open access journal *Micromachines* (ISSN 2072-666X) form 2018 to 2019 (available at: https://www.mdpi.com/journal/micromachines/special_issues/Nanostructure_based_Sensors_Gas_Sensing)

For citation purposes, cite each article independently as indicated on the article page online and as indicated below:

LastName, A.A.; LastName, B.B.; LastName, C.C. Article Title. *Journal Name* **Year**, *Article Number*, Page Range.

ISBN 978-3-03921-636-9 (Pbk)
ISBN 978-3-03921-637-6 (PDF)

# Contents

# About the Special Issue Editors

**Nicola Donato** received the M.S. degree in Electronic Engineering from the University of Messina, Messina, Italy, and the Ph.D. degree from the University of Palermo, Palermo, Italy. He is currently an Associate Professor of Electrical and Electronic Measurements and the head of the laboratories of "Electronics for Sensors and for Systems of Transduction" and of "Electrical and Electronic Measurements" at University of Messina. He has authored over 150 papers on international journals and conference proceedings (Scopus). His current research interests include sensor characterization and modeling, development of measurement system for sensors, and characterization of electronic devices up to microwave range and downto criogenic temperatures.

**Sabrina Grassini** received the M.S. degree in chemistry from the University of Torino, Turin, Italy, in 1999, and the Ph.D. degree in metallurgical engineering from the Politecnico di Torino, Turin, in 2004. In 2007, she joined the Politecnico di Torino as an Assistant Professor of chemistry and became an Associate Professor of applied physical chemistry with the Department of Applied Science and Technology in 2016. Her current research interests include chemical/physical fundamentals of plasma processes, the study of corrosion mechanisms of metals and alloys, conservation of cultural heritage, and sensors for environmental and medical measurements. She has authored more than 180 papers in national and international journals and in the proceedings of international conferences.

*micromachines*

**MDPI**

*Editorial*

# Editorial for the Special Issue on Nanostructure Based Sensors for Gas Sensing: from Devices to Systems

**Nicola Donato** [1,*] **and Sabrina Grassini** [2,*]

1 Department of Engineering, University of Messina, 98122 Messina, Italy
2 Department of Applied Science and Technology Politecnico di Torino, 10129 Torino, Italy
* Correspondence: ndonato@unime.it (N.D.); sabrina.grassini@polito.it (S.G.)

Received: 12 August 2019; Accepted: 4 September 2019; Published: 9 September 2019

The development of solid state gas sensors based on microtransducers and nanostructured sensing materials is the key point in the design of new portable measurement systems with sensing and identification performances comparable with those of most sophisticated analytical techniques. In such a context, a lot of effort must be spent of course in the development of the sensing material, but also in the choice of the transducer mechanism and structure, in the electrical characterization of the sensor prototypes, as well as in the design of suitable measurement setups.

After a careful peer review, seven manuscripts covering all the aspects of the sensor world were accepted for publication in this special issue. Papers [1,2] deal with sensing material preparation and the characterization of the chemico-physical and sensing properties, while further studies report about the investigation of sensing performance towards different operating conditions [3] and the optimization of the transduction mechanism and of the device package [4]. Furthermore, there are three papers focused on gas sensor systems and their application in environmental monitoring [5,6] and in the biomedical field [7].

In more detail, Xu and co-authors describe a route to fabricate gold nanoparticles (less than 20 nm in diameter) wrapped with a controllable ultrathin carbon layer (Au@C, 0.6–2 nm thick) by one step laser ablation of the noble metal target in toluene–ethanol mixed solutions. The developed sensing material was tested for the detection of low concentrations of $H_2S$ gas, ranging from 1 to 5 ppm, at room temperature [1].

Li et al. present the electrospray process to deposit ZnO patterns for gas sensing, paying particular attention on the effects of different experimental parameters on the jet characteristics and on the final properties of electrosprayed patterns. Sensing performance towards alcohol vapors are also well discussed [2].

Bonaccorsi et al. show how UV irradiation can improve the response of an indium oxide ($In_2O_3$) resistive sensor to detect carbon monoxide, operating at low temperature in the range of 25–150 °C. In particular, the best balance between operating temperature and UV irradiation toward low CO concentration values (from 1 to 10 ppm) was observed at 100 °C [3].

Considering the investigation of transduction mechanism and packaging steps, Yildiz et al. present the fabrication and packaging of a capacitive micromachined ultrasonic transducer (CMUT) using anodically bondable low temperature co-fired ceramic (LTCC). The authors point out the attention on a promising approach for high density CMUT array fabrication and the indirect integration of CMUT-IC for a miniature size packaging [4].

Micromachining technology is the new frontier in the realization of miniaturized systems; as a matter of fact, Jianhai Sun et al. developed a mini monitoring system integrated with a microfabricated metal oxide array sensor and a micro packed gas chromatographic (GC) column for detecting environmental gases [5]. By using the chromatographic separation capability, the MOS array sensor was able to detect each component, avoiding the technical bottleneck of mutual interference among different gases.

*Micromachines* **2019**, *10*, 591

Mao et al. present a set of hardware platforms to improve the efficiency of new developed E-nose; the proposed system includes a gas-sensing, film-parallel, synthesis platform, a high-throughput gas sensing unmanned testing platform, and a handheld E-nose system. The sensor arrays are produced by inkjet printing, tailoring the devices for the specific application [6].

Ultimately, in the biomedical field, the design and development of mini-invasive systems for gas monitoring is a real challenge. In such a scenario, breath analysis is one of the best candidates, so Gatty et al. developed and characterized an integrated amperometric sensor [7] in order to determine the hydrogen sulphide ($H_2S$) concentration, one of the main reasons of malodour, in oral breath.

We hope that this special issue gives the reader new points of view in gas sensing and miniaturized systems, taking into account their fundamental role in environmental safety and human health. The special issue wants to highlight the importance of synergy among micromachining, instrumentation and measurement, chemistry and material science to face needs and challenges in gas sensor design and development.

We would like to take this opportunity to thank all the authors for submitting their papers to this special issue. We would like to thank also all reviewers for their efforts and comments to improve the quality of the submitted papers.

**Conflicts of Interest:** The authors declare no conflict of interest.

## References

1. Xu, X.; Gao, L.; Duan, G. The Fabrication of Au@C Core/Shell Nanoparticles by Laser Ablation in Solutions and Their Enhancements to a Gas Sensor. *Micromachines* **2018**, *9*, 278. [CrossRef] [PubMed]
2. Li, W.; Lin, J.; Wang, X.; Jiang, J.; Guo, S.; Zheng, G. Electrospray Deposition of ZnO Thin Films and Its Application to Gas Sensors. *Micromachines* **2018**, *9*, 66. [CrossRef] [PubMed]
3. Bonaccorsi, L.; Malara, A.; Donato, A.; Donato, N.; Leonardi, S.G.; Neri, G. Effects of UV Irradiation on the Sensing Properties of $In_2O_3$ for CO Detection at Low Temperature. *Micromachines* **2019**, *10*, 338. [CrossRef] [PubMed]
4. Yildiz, F.; Matsunaga, T.; Haga, Y. Fabrication and Packaging of CMUT Using Low Temperature Co-Fired Ceramic. *Micromachines* **2018**, *9*, 553. [CrossRef] [PubMed]
5. Sun, J.; Geng, Z.; Xue, N.; Liu, C.; Ma, T. A Mini-System Integrated with Metal-Oxide-Semiconductor Sensor and Micro-Packed Gas Chromatographic Column. *Micromachines* **2018**, *9*, 408. [CrossRef] [PubMed]
6. Mao, Z.; Wang, J.; Gong, Y.; Yang, H.; Zhang, S. A Set of Platforms with Combinatorial and High-Throughput Technique for Gas Sensing, from Material to Device and to System. *Micromachines* **2018**, *9*, 606. [CrossRef] [PubMed]
7. Gatty, H.K.; Stemme, G.; Roxhed, N. A Miniaturized Amperometric Hydrogen Sulfide Sensor Applicable for Bad Breath Monitoring. *Micromachines* **2018**, *9*, 612. [CrossRef] [PubMed]

*micromachines*

MDPI

*Article*

# Effects of UV Irradiation on the Sensing Properties of In₂O₃ for CO Detection at Low Temperature

Lucio Bonaccorsi [1,*], Angela Malara [1], Andrea Donato [1], Nicola Donato [2,*], Salvatore Gianluca Leonardi [2] and Giovanni Neri [2]

[1] Dipartimento DICEAM, Università Mediterranea, Loc. Feo di Vito, 89060 Reggio Cal, Italy; angela.malara@unirc.it (A.M.); andrea.donato@unirc.it (A.D.)
[2] Dipartimento di Ingegneria, Università di Messina, C.da Di Dio, 98166 Messina, Italy; leonardis@unime.it (S.G.L.); gneri@unime.it (G.N.)
* Correspondence: lucio.bonaccorsi@unirc.it (L.B.); ndonato@unime.it (N.D.)

Received: 29 March 2019; Accepted: 17 May 2019; Published: 22 May 2019

**Abstract:** In this study, UV irradiation was used to improve the response of indium oxide (In₂O₃) used as a CO sensing material for a resistive sensor operating in a low temperature range, from 25 °C to 150 °C. Different experimental conditions have been compared, varying UV irradiation mode and sensor operating temperature. Results demonstrated that operating the sensor under continuous UV radiation did not improve the response to target gas. The most advantageous condition was obtained when the UV LED irradiated the sensor in regeneration and was turned off during CO detection. In this operating mode, the semiconductor layer showed an apparent "p-type" behavior due to the UV irradiation. Overall, the effect was an improvement of the indium oxide response at 100 °C toward low CO concentrations (from 1 to 10 ppm) that showed higher results than in the dark, which is promising to extend the detection of CO with an In₂O₃-based sensor in the sub-ppm range.

**Keywords:** indium oxide; UV irradiation; CO detection

## 1. Introduction

Indium oxide is a semiconductor metal oxide (MOX) which displays very good performances when used as a sensing layer in resistive gas sensors for the detection of oxidizing gases like $O_3$ and $NO_2$ [1–5]. In₂O₃ has been investigated less for CO detection at very low concentrations. Further, previous In₂O₃-based CO sensors needed to be operated at elevated temperatures, in the range of 250–400 °C [6–9]. In a previous paper, we reported a fast and repeatable response toward this gas, allowing the detection of less than 2 ppm of CO in a few seconds, operating the sensor at 250 °C [10]. As the working temperature is one of the important parameters that determines the effective use of a sensor in commercial applications, the development of sensors for the monitoring of very low CO concentrations at low temperature is presently an active research field because these devices have lower operating costs, longer lifespans and an increased stability [11,12].

To improve performances of resistive In₂O₃ sensors at lower temperatures, different routes are possible. Doping indium oxide with low concentrations of additives has been proved to be beneficial in promoting sensitivity and lowering operating temperatures [2,5–8]. UV irradiation has been also successfully reported in recent papers for improving the performances of the In₂O₃ sensing surface [13–17]. Indium oxide, indeed, is a UV responsive metal oxide since UV photons radiation causes the formation of electron/hole pairs in the depletion region of the oxide grains that increase the intra-grain conductivity of the sensing layer [18]. In detection of oxidizing gases such as $NO_2$, it is well known that the UV radiation of the semiconductor oxide surface causes larger variations of the electrical resistance, that in turn improves the sensing performance of In₂O₃ at lower temperatures [6,16,19].

However, very little data is available about the effect of UV in promoting CO sensing on $In_2O_3$ and the related phenomena involved [13]. It is known, indeed, that UV illumination has multiple effects on oxygen adsorption, conductance of the sensing layer and CO adsorption/reactivity [13,15,17]. In this work, we report the results of an investigation on CO sensing properties of $In_2O_3$ under different UV irradiation modes to improve MOX response at low temperatures. CO concentrations in the range of 1–10 ppm and sensor temperatures from 25 °C to 150 °C were used in combination with UV light activated during CO detection or not. Results showed that UV light generally did not improve $In_2O_3$ response at low temperatures, however, when used in certain conditions, the UV photons irradiation demonstrated to increase the sensor response. This demonstrates the possibility to detect sub-ppm concentrations of CO with $In_2O_3$-based sensor at a relatively low temperature without any use of additives.

## 2. Materials and Methods

### 2.1. Sample Preparation and Characterization

Indium oxide powder was prepared by precipitation from an aqueous solution of indium nitrate (0.68 M) hydrolyzed with an aqueous potassium carbonate solution (1 M). The obtained precipitate was filtered, washed with deionized water, dried at 110 °C for 12 h and then calcined at 500 °C for 12 h in air [7].

Powder sample was characterized by X-ray powder diffraction (XRD) analysis (Bruker, D2 Phaser, Karlsruhe, Germany) in the 2θ range from 10 to 80° (Cu Kα1 = 1.54056 Å) and its morphology studied by Scanning Electron Microscopy SEM (Phenom ProX, Deben, Suffolk, UK). The Brunauer–Emmett–Teller (BET) surface areas and the total Pore Volume of the prepared $In_2O_3$ powder were determined from nitrogen adsorption–desorption isotherms at 77 K (ChemiSorb 2750 Micromeritics, Norcross, GA, USA).

### 2.2. Sensor Preparation and Testing

Sensors were prepared by depositing a paste of indium oxide powder mixed with a proper quantity of ethanol onto an alumina planar substrate (3 mm × 6 mm) supplied with interdigitated Pt electrodes and a heating element on the back side. Before sensing tests, the sensors were conditioned in air for 2 h at 400 °C to stabilize the deposited film. Measurements were performed positioning the sensor in a testing cell and flowing a mixture of dry air and CO at different concentrations for a total gas stream of 100 sccm. All gas fluxes were measured by computer-controlled mass flow meters. The sensors resistance data were collected in the four-point mode by an Agilent 34970A (Santa Clara, CA, USA) multimeter while a dual-channel power supplier instrument (Agilent E3632A, Santa Clara, CA, USA) allowed controlling the sensor temperature.

The testing cell was equipped with an UV LED (λ = 400 nm, I = 20 mW/cm$^2$) sited in front of the sensing layer of the sensor. The UV irradiation of the sensor surface was varied during the experiments, according to the diagram shown in Figure 1. Three different CO concentrations and sensor temperatures have been tested using four different UV irradiation modes for each case: UV (always) Off, UV (always) On, UV On in sensor regeneration (airflow), UV On in detection (air+CO) (Figure 1).

| UV led status | | | |
|---|---|---|---|
| UV off | UV on | UV on/air | UV on/air+CO |

| T sensor (°C) | | |
|---|---|---|
| 25 | 100 | 150 |

| CO concentration (ppm) | | |
|---|---|---|
| 1 | 5 | 10 |

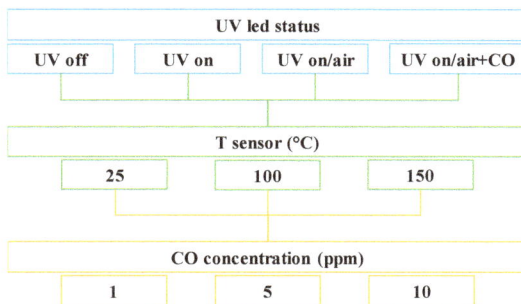

**Figure 1.** Diagram showing the experimental conditions used.

## 3. Results and Discussion

### 3.1. Morphological and Microstructural Characterization

The morphology of the synthesized $In_2O_3$ powder was investigated by SEM analysis. The image in Figure 2 shows that the $In_2O_3$ powder is constituted by small particles with a characteristic cubic shape, assembled in larger agglomerate with wide size distribution ranging from 0.1 to 0.8 µm.

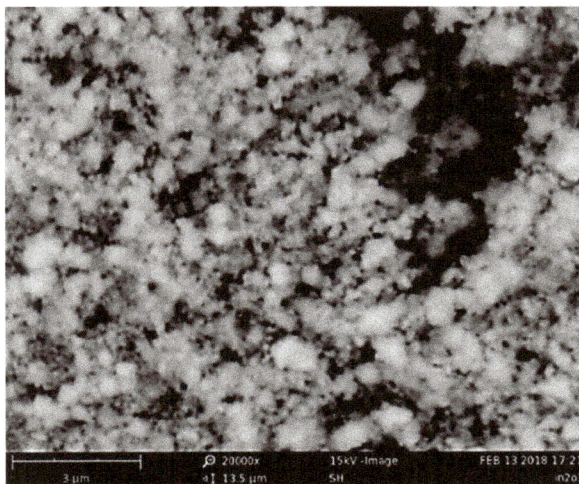

**Figure 2.** Scanning Electron Microscopy (SEM) image of the prepared $In_2O_3$.

XRD analysis demonstrated that these particles are highly crystalline. The XRD spectrum of precipitated powder after calcination (T = 500 °C) confirmed the formation of crystalline indium (III) oxide in the typical cubic phase, bixbyite (Figure 3). The crystallites size calculated by the Scherrer formula from diffraction peaks in Figure 3 resulted ~30 nm, confirming that the larger granules observed by electron microscopy are aggregates.

The above morphological and microstructural characteristics suggest that this sample likely has a large surface area. In fact, the BET surface area of the precipitated indium oxide, measured by nitrogen adsorption method, was 20.3 $m^2/g$ and the total Pore Volume = 87 $mm^3/g$. For comparison, the surface area of a commercial $In_2O_3$ (Sigma Aldrich) resulted much lower: only 10.2 $m^2/g$ with a pore volume of 42 $mm^3/g$.

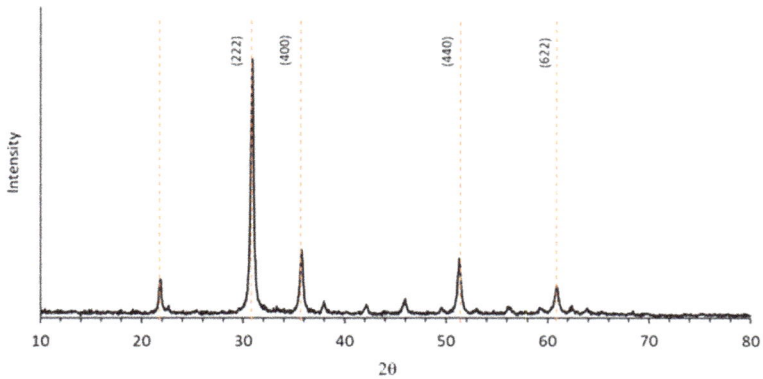

**Figure 3.** XR diffractogram of the calcined indium oxide powder.

*3.2. Gas Sensor Measurement*

Indium oxide is a MOX semiconductor showing typical n-type behavior, i.e., the electrical resistance of the deposited layer decreases in presence of CO due to the oxidation of the reducing target gas on the MOX surface. Measurements carried out in dark conditions (see Figures 4 and 5) confirm this behavior.

Figure 4 shows the sensor resistance variation with temperature for a concentration of 5 ppm CO. Although the temperatures range considered was low, the baseline resistance in the airflow was lower than 1 k$\Omega$ even at T = 25 °C and decreased increasing the sensor temperature to 100 and 150 °C (Figure 4). The low values shown by In$_2$O$_3$ are due to the high intrinsic electron concentration with good mobility in the sensing layer of this semiconductor metal oxide [20,21].

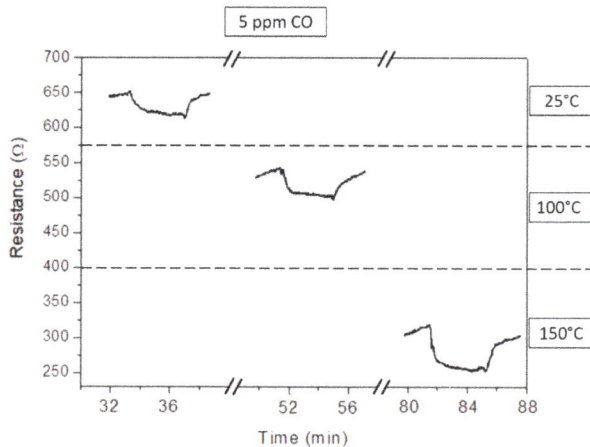

**Figure 4.** Transient response of the indium oxide sensor to 5 ppm CO at different operating temperatures.

The transient response at T = 25 °C and 100 °C was weak but showed a complete and reversible recovery in few minutes at the removal of the target gas (Figure 4). Increasing the temperature to 150 °C, the recovery time to the baseline was similar although the resistance variation increased. Comparable values were observed at the different CO concentrations, as shown by the sensor transient response at T = 100 °C in Figure 5. The recovery time of the In$_2$O$_3$ layer was not influenced by the CO concentration, showing that the adsorption/desorption and reaction processes occurring on the sensor surface were unaffected.

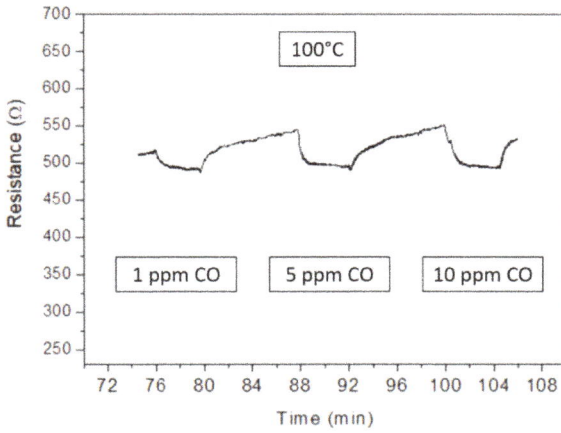

**Figure 5.** Transient sensor response at T = 100 °C and different CO concentrations.

To evidence the sensor behavior, the response $S = R_0/R$, where $R_0$ is the baseline resistance in air and R is the resistance at different CO concentrations, is plotted for the three tested temperatures in Figure 6. In "UV Off" mode (Figure 6a), the response at room temperature was very low, however it increased significantly by heating the sensor surface at 100 °C and even more at 150 °C.

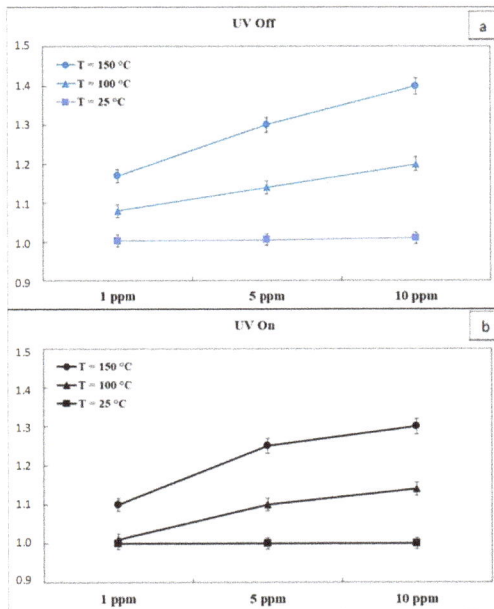

**Figure 6.** In$_2$O$_3$ sensor response at different CO concentrations and temperatures in "UV Off" (a) and "UV On" (b) mode. Error bars are calculated on the average of 5 runs.

The ideal working temperature for resistive sensors is generally higher than 200 °C [6–10] and considering the low temperatures and the CO concentrations range tested, the $In_2O_3$ sensor has however shown a weak but detectable sensitivity.

In the same conditions, the continuous irradiation of the sensor surface, i.e., in the "UV On" mode (Figure 6b), was never beneficial. Indeed, comparing data in Figure 6, the sensor response in "UV On" was similar or even lower than in dark mode, for all the tested temperatures. It is of interest to observe that increasing the sensor surface temperature, the response in dark mode increased almost linearly from 1 ppm to 10 ppm of CO (Figure 6a) while under UV irradiation (Figure 6b) the response increase at 150 °C was clearly not linear. The observed behavior was ascribed to the combined effect of the UV photo-activation and the sensor temperature that favored the CO desorption from the sensing surface, as explained in the following discussion (Section 3.3).

The most interesting results were obtained when the sensor surface was irradiated with UV in a non-continuous mode, i.e., "UV On in air" mode. Figure 7 shows the variation of the sensor layer resistance vs. time at the operating temperature of 100 °C, under CO pulses of different concentration. The UV irradiation of the sensor surface in airflow caused a resistance decrease that reverted when the UV LED was switched off and the sensor was exposed to the CO pulse (Figure 7). Increasing CO in air, from 1 ppm to 10 ppm, the observed resistance variation increased showing a correlation with the CO concentration.

It was clearly noted that, when the UV LED was switched on during the sensor surface regeneration, the sensor resistance in CO, $R_{CO}$, was higher than the resistance in air suggesting a p-type behavior (Figure 7).

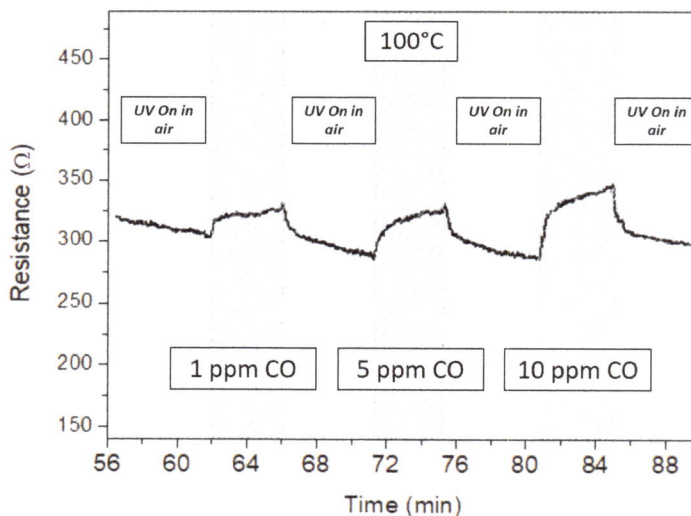

**Figure 7.** Transient sensor response at T = 100 °C and different CO concentrations in "UV On in air" mode.

In Figure 8a–c are summarized data acquired in all the above described conditions. In each plot of Figure 8 is shown the sensor response to a CO pulse varying the sensor temperature in the three tested UV modes. From data in Figure 8a–c, the room temperature was always a condition of low response for the $In_2O_3$ sensor while for higher temperatures, 100 °C and 150 °C, the CO detection was clearly observed.

As pointed out before, UV irradiation during CO detection, "UV On in air+CO", was a condition that did not show a real advantage in terms of sensor response compared to the "UV Off" mode, for all the operating temperatures tested (Figure 8a–c).

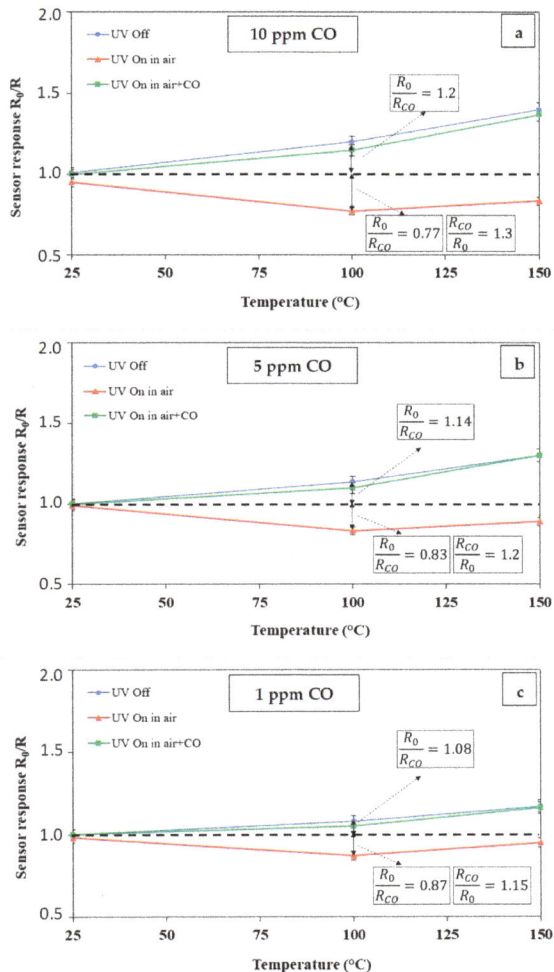

**Figure 8.** Sensor response to (**a**) 10 ppm; (**b**) 5 ppm and (**c**) 1 ppm of CO at different operating temperatures and different UV irradiation modes. Error bars are calculated on the average of five runs.

Irradiating the sensor surface in the airflow at T = 100 °C and 150 °C, according to curves in Figure 8a–c, the sensor response was <1, like in a p-type semiconductor, where the sensor resistance in CO, $R_{CO}$, resulted higher than the resistance in air, $R_0$. For a CO concentration of 10 ppm and T = 100 °C it was (Figure 8a):

$$\left(\frac{R_0}{R_{CO}}\right)_{UV\ Off} = 1.2 \text{ and } \left(\frac{R_0}{R_{CO}}\right)_{UV\ On\ in\ Air} = 0.77$$

However, inverting the previous ratio (Figure 8a):

$$\left(\frac{R_{CO}}{R_0}\right)_{\text{UV On in Air}} = 1.3$$

By these results, it appears that the right combination of sensor temperature and UV light operation mode is capable of increasing the sensor performance towards CO detection. In the specific, at the operating temperature of 100 °C, the p-type response of $In_2O_3$ in "UV On in air" mode is higher than in dark or in "UV On in air+CO" for all the tested CO concentrations. The advantage of irradiating the sensor only during the regeneration in airflow, however, was lost when the operating temperature was increased to 150 °C.

Motivated by the interesting features displayed, we further studied the performances of the sensor in these optimized conditions. The sensitivity of the sensors was then evaluated plotting the sensor responses vs. CO concentration (Figure 9a) for assessing its suitability in the monitoring of CO. The sensor showed good response to the smallest tested concentration of CO (1 ppm).

When the responses are plotted in a log–log scale, a linear trend was observed as a function of the gas target concentration (Figure 9b). The limit of detection, defined as the lower concentration at which the response is significantly differentiated from the noise signal (usually at S/N = 3), was extrapolated to be below 200 ppb. This very low detection limit suggests that the photoactivated $In_2O_3$-based sensor can be applied for the practical use in the field of environmental control.

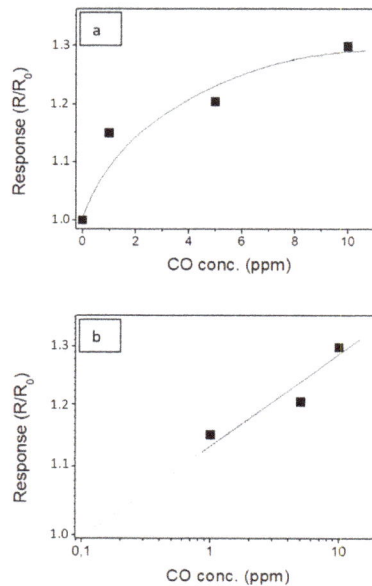

**Figure 9.** (**a**) Calibration curve showing the sensor response to different CO concentrations under the "UV On in air" method; (**b**) calibration curve plotted in log-log scale.

### 3.3. Sensing Mechanism

Based on results above reported, a discussion on the CO sensing mechanism of the sensor proposed is presented. Indium oxide is a semiconductor with a band gap of 3.5–3.7 eV that shows n-type behavior due to intrinsic defects (oxygen vacancies) [13,22,23]. UV irradiation is generally considered to improve the sensor response because the photons radiation of the semiconductor surface changes the surface potential at the grain boundaries [13,23] enhancing the conductance of the sensing layer [16,17]. In this sense, the UV irradiation can reduce the operating temperature of the metal oxide allowing, in some cases, the use of the resistive sensor at room temperature [15]. In our experiments, however, the continuous UV irradiation of the In$_2$O$_3$ sensor did not result in an improved response. According to the study of Espid et al. [17], a high UV irradiation of the sensing surface increases the generation rate of electron/hole pairs but also their likelihood of recombination thus favoring the desorption of the molecular and atomic oxygen interacting with the charge carriers UV generated:

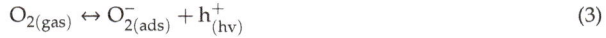

$$0 \rightarrow e^-_{(h\nu)} + h^+_{(h\nu)} \tag{1}$$

$$O_{2(gas)} + e^-_{(h\nu)} \leftrightarrow O^-_{2(ads)} \tag{2}$$

$$O_{2(gas)} \leftrightarrow O^-_{2(ads)} + h^+_{(h\nu)} \tag{3}$$

The final effect was that under continuous UV light, the ratio $R_0/R_{air+CO}$ was similar or even lower than with the UV LED off.

For a similar reason, the UV photo-activation during CO oxidation, "UV On in Air+CO" mode, although caused the lowering of the baseline resistance $R_0$, did not improve the sensor response due to the rapid desorption of oxygen even at low temperatures.

The improved sensor response obtained in "UV On in Air" mode is related to the "p-type" behavior shown by the indium oxide layer when irradiated in the airflow. In Figure 10, the baseline resistance at T = 100 °C in UV Off mode, $R_0$ = 550 Ω, decreased to $R'_0$ = 280 Ω when the UV LED was switched on in the airflow and then increased to $R_{CO}$ = 360 Ω when the UV was off and the sensor exposed to a 10 ppm CO pulse.

The significant decrease of the sensor baseline resistance mainly due to the generation of a large number of electron/hole pairs in the semiconductor particles was followed by an increase in resistance as soon as the UV LED was switched off and the CO + air mixture was sent.

The charge carriers recombination was the dominant effect in terms of sensor resistance variation prevailing the contemporaneous charges addition due to the CO oxidation:

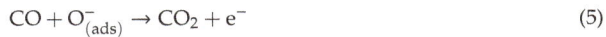

$$e^-_{(h\nu)} + h^+_{(h\nu)} \rightarrow 0 \tag{4}$$

$$CO + O^-_{(ads)} \rightarrow CO_2 + e^- \tag{5}$$

The combination of the two opposite effects causes that the sensor resistance does not return to the baseline value $R_0$ but stops to a value that is proportional to the CO concentration. As shown in Figure 10, the observed p-type behavior was not, in fact, due to a real change in the semiconductor properties of In$_2$O$_3$ but resulting from the combination of UV illumination and operating conditions.

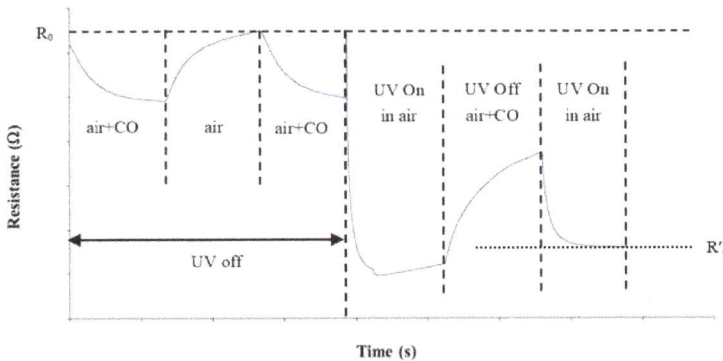

**Figure 10.** Transient sensor response in "UV Off" mode and "UV On in air" mode. (T = 100 °C, 10 ppm CO).

At the sensor temperature of 150 °C, however, the positive effects of working in the "UV On in air" mode decreased. By increasing the temperature of the MOX semiconductors, different effects are produced, such as the modification of the surface potential of the sensitive layer, the adsorbed oxygen species and their adsorption/desorption rate, the speed of the chemical reactions of the molecules/atoms involved [11,12]. The worsening of the sensor response observed at 150 °C in UV light in regeneration, however, was ascribed to the increase of the recombination rate of the UV-generated charges with temperature [18,24] and the consequent decrease of the ratio $R_{air+CO}/R_0$.

## 4. Conclusions

The effect of the UV light on the sensing properties of indium oxide in CO detection at low temperatures has been studied by varying the irradiation mode of the sensing layer. Results showed that operating the sensor under UV radiation did not improve the $In_2O_3$ response at low temperatures in comparison to the typical working condition in dark. An improvement was observed when the UV LED irradiated the sensing layer only in airflow, namely in the regeneration phase, and was turned off in detection. The effect, however, was particularly significant at the sensor temperature of 100 °C. In these operating conditions, the enhancement of response observed could be exploited for detecting sub-ppm concentrations of CO (up to 200 ppb) at a relatively low temperature, suggesting that the photoactivated $In_2O_3$ sensor can be applied for the practical use in the field of environmental control.

**Author Contributions:** Conceptualization: L.B. and N.D.; investigation: A.M.; validation: S.G.L.; writing—original draft preparation: A.M.; writing—review and editing: S.G.L. and A.M.; supervision and project administration: A.D. and G.N.

**Funding:** This research received no external funding.

**Conflicts of Interest:** The authors declare no conflict of interest.

## References

1. Neri, G. First Fifty Years of Chemoresistive Gas Sensors. *Chemosensors* **2015**, *3*, 1–20. [CrossRef]
2. Forleo, A.; Francioso, L.; Epifani, M.; Capone, S.; Taurino, A.M.; Siciliano, P. NO₂- gas-sensing properties of mixed $In_2O_3$–$SnO_2$ thin films. *Thin Solid Films* **2005**, *490*, 68–73. [CrossRef]
3. Korotcenkov, G.; Brinzari, V.; Cerneavschi, A.; Ivanov, M.; Golovanov, V.; Cornet, A.; Morante, J.; Cabot, A.; Arbiol, J. The influence of film structure on $In_2O_3$ gas response. *Thin Solid Films* **2004**, *460*, 315–323. [CrossRef]
4. Chung, W.Y.; Sakai, G.; Shimanoe, K.; Miura, N.; Lee, D.D.; Yamazoe, N. Preparation of indium oxide thin film by spin-coating method and its gas-sensing properties. *Sens. Actuators B Chem.* **1998**, *46*, 139–145. [CrossRef]
5. Korotcenkov, G.; Boris, I.; Cornet, A.; Rodriguez, J.; Cirera, A.; Golovanov, V.; Lychkovsky, Y.; Karkotsky, G. The influence of additives on gas sensing and structural properties of $In_2O_3$-based ceramics. *Sens. Actuators B Chem.* **2007**, *120*, 657–664. [CrossRef]
6. Ivanovskaya, M.; Bogdanov, P.; Faglia, G.; Sberveglieri, G. The features of thin film and ceramic sensors at the detection of CO and NO₂. *Sens. Actuators B Chem.* **2000**, *68*, 344–350. [CrossRef]
7. Yamaura, H.; Jinkawa, T.; Tamaki, J.; Moriya, K.; Miura, N.; Yamazoe, N. Indium oxide-based gas sensor for selective detection of CO. *Sens. Actuators B Chem.* **1996**, *35–36*, 325–332. [CrossRef]
8. Yamaura, H.; Moriya, K.; Miura, N.; Yamazoe, N. Mechanism of sensitivity promotion in CO sensor using indium oxide and cobalt oxide. *Sens. Actuators B Chem.* **2000**, *65*, 39–41. [CrossRef]
9. Lim, S.K.; Hwang, S.H.; Chang, D.; Kim, S. Preparation of mesoporous $In_2O_3$ nanofibers by electrospinning and their application as a CO gas sensor. *Sens. Actuators B Chem.* **2010**, *149*, 28–33. [CrossRef]
10. Donato, N.; Neri, F.; Neri, G.; Latino, M.; Barreca, F.; Spadaro, S.; Pisagatti, I.; Currò, G. CO sensing devices based on indium oxide nanoparticles prepared by laser ablation in water. *Thin Solid Films* **2011**, *520*, 922–926. [CrossRef]
11. Dey, A. Semiconductor metal oxide gas sensors: A review. *Mater. Sci. Eng. B* **2018**, *229*, 206–217. [CrossRef]
12. Korotcenkov, G. Metal oxides for solid-state gas sensors: What determines our choice? *Mater. Sci. Eng. B* **2007**, *139*, 1–23. [CrossRef]

13. Comini, E.; Cristalli, A.; Faglia, G.; Sberveglieri, G. Light enhanced gas sensing properties of indium oxide and tin dioxide sensors. *Sens. Actuators B Chem.* **2000**, *65*, 260–263. [CrossRef]

14. Barquinha, P.; Pimentel, A.; Marques, A.; Pereira, L.; Martins, R.; Fortunato, E. Effect of UV and visible light radiation on the electrical performances of transparent TFTs based on amorphous indium zinc oxide. *J. Non. Cryst. Solids* **2006**, *352*, 1756–1760. [CrossRef]

15. Wang, C.Y.; Becker, R.W.; Passow, T.; Pletschen, W.; Köhler, K.; Cimalla, V.; Ambacher, O. Photon stimulated sensor based on indium oxide nanoparticles I: Wide-concentration-range ozone monitoring in air. *Sens. Actuators B Chem.* **2011**, *152*, 235–240. [CrossRef]

16. Wagner, T.; Kohl, C.D.; Malagù, C.; Donato, N.; Latino, M.; Neri, G.; Tiemann, M. UV light-enhanced $NO_2$ sensing by mesoporous $In_2O_3$: Interpretation of results by a new sensing model. *Sens. Actuators B Chem.* **2013**, *187*, 488–494. [CrossRef]

17. Espid, E.; Taghipour, F. Development of highly sensitive $ZnO/In_2O_3$ composite gas sensor activated by UV-LED. *Sens. Actuators B Chem.* **2017**, *241*, 828–839. [CrossRef]

18. Mishra, S.; Ghanshyam, C.; Ram, N.; Bajpai, R.P.; Bedi, R.K. Detection mechanism of metal oxide gas sensor under UV radiation. *Sens. Actuators B Chem.* **2004**, *97*, 387–390. [CrossRef]

19. Wagner, T.; Kohl, C.D.; Morandi, S.; Malagù, C.; Donato, N.; Latino, M.; Neri, G.; Tiemann, M. Photoreduction of Mesoporous $In_2O_3$: Mechanistic Model and Utility in Gas Sensing. *Chem. Eur. J.* **2012**, *18*, 8216–8223. [CrossRef]

20. Neri, G.; Bonavita, A.; Micali, G.; Rizzo, G.; Callone, E.; Carturan, G. Resistive CO gas sensors based on $In_2O_3$ and $InSnO_x$ nanopowders synthesized via starch-aided sol-gel process for automotive applications. *Sens. Actuators B Chem.* **2008**, *132*, 224–233. [CrossRef]

21. Yamazoe, N.; Sakai, G.; Shimanoe, K. Oxide semiconductor gas sensors. *Catal. Surv. Asia* **2003**, *7*, 63–75. [CrossRef]

22. Bender, M.; Katsarakis, N.; Gagaoudakis, E.; Hourdakis, E.; Douloufakis, E.; Cimalla, V.; Kiriakidis, G. Dependence of the photoreduction and oxidation behavior of indium oxide films on substrate temperature and film thickness. *J. Appl. Phys.* **2001**, *90*, 5382. [CrossRef]

23. Werner, R.L.; Ley, R.P. Optical Properties of Indium Oxide. *J. Appl. Phys.* **1966**, *37*, 299.

24. De Lacy Costello, B.P.J.; Ewen, R.J.; Ratcliffe, N.M.; Richards, M. Highly sensitive room temperature sensors based on the UV-LED activation of zinc oxide nanoparticles. *Sens. Actuators B Chem.* **2008**, *134*, 945–995. [CrossRef]

*micromachines*

MDPI

*Article*

# A Miniaturized Amperometric Hydrogen Sulfide Sensor Applicable for Bad Breath Monitoring

**Hithesh K. Gatty, Göran Stemme and Niclas Roxhed ***

Micro and Nanosystems, KTH Royal Institute of Technology, SE-100 44 Stockholm, Sweden;
hithesh@kth.se (H.K.G.); stemme@kth.se (G.S.)
* Correspondence: roxhed@kth.se

Received: 2 October 2018; Accepted: 14 November 2018; Published: 22 November 2018

**Abstract:** Bad breath or halitosis affects a majority of the population from time to time, causing personal discomfort and social embarrassment. Here, we report on a miniaturized, microelectromechanical systems (MEMS)-based, amperometric hydrogen sulfide ($H_2S$) sensor that potentially allows bad breath quantification through a small handheld device. The sensor is designed to detect $H_2S$ gas in the order of parts-per-billion (ppb) and has a measured sensitivity of 0.65 nA/ppb with a response time of 21 s. The sensor was found to be selective to NO and $NH_3$ gases, which are normally present in the oral breath of adults. The ppb-level detection capability of the integrated sensor, combined with its relatively fast response and high sensitivity to $H_2S$, makes the sensor potentially applicable for oral breath monitoring.

**Keywords:** hydrogen sulfide; amperometric; MEMS; gas sensor; bad breath; halitosis

## 1. Introduction

Bad breath or oral malodor, affects a majority of the population on a regular basis. The presence of plaque, tongue coating [1], gum diseases [2], exposed necrotic tooth pulp, and healing wounds [3] are known to be the cause of oral malodor. Microorganisms present in oral cavities react with organic compounds, releasing sulfur-containing by-products that lead to bad breath. Specifically, sulfur-containing by-products, such as hydrogen sulfide ($H_2S$), methyl mercaptan ($CH_4S$), and dimethyl sulfide (($CH_3)_2S$), are associated with bad breath, which is also termed as halitosis [4].

Until recently, oral malodor was diagnosed by physicians in a purely subjective manner (smelling). However, recent developments in sensor technology have provided measuring instruments with sensitive detection of bad breath. The most successful commercial measuring instrument is the Halimeter™ [5], a standard clinical bench-top instrument used to measure volatile sulfur compounds (VSCs), particularly $H_2S$ gas concentration. In this instrument, the user blows into a tube attached to the instrument and a concentration value is presented on a display. Halitosis in an adult is classified as "normal" if the concentration is within the range of 80–160 parts-per-billion (ppb), "weak" if the concentration is within the range of 160–250 ppb, and "strong" if the concentration is greater than 250 ppb [6,7]. The disadvantage of the Halimeter instrument is that it is a bench-top apparatus (3.6 kg) that requires warm-up times and yearly maintenance [8] and is thus an instrument that is primarily designed for patient examination or population studies. However, to more directly address and counteract personal discomfort, ad-hoc mobile monitoring of bad breath would be highly desired. To achieve such monitoring, the sensor element is essential, which requires a small form factor for integration, a fast response time, and ppb-level sensitivity.

Among the various types of $H_2S$ sensors developed, amperometric sensors are particularly advantageous as it allows the fabrication of miniaturized and high sensitivity sensors with fast response time. Schiavon and Zotti achieved detection limits of 45 ppb using discrete porous silver electrodes supported on separate ion-exchange membranes [9]. However, nonintegrated discrete components result in relatively large-size sensors, which is undesirable when developing a handheld instrument. Recently, Yang et al. showed a fast response Nafion-based amperometric sensor that could detect $H_2S$ in the range of 0.1–200 ppm [10]. However, a complex fabrication method of the sensing electrode and lower sensitivity limits the sensor from being used in the ppb range, which is required for bad breath detection.

In the present work, a miniaturized and integrated electrochemical $H_2S$ sensor with fast response time and a ppb-level sensitivity that is applicable for Halitosis measurement is demonstrated. A simple fabrication method involving high aspect ratio etching and atomic layer deposition of platinum provides the basic structure for preparation of the sensing electrode. The sensor was characterized for its cross sensitivity to nitric oxide (NO), which is normally present in the oral cavity and nasal cavity. Nasal cavity NO contributes to the high concentration in the oral cavity and can affect the NO concentration in the oral cavity. A typical concentration of NO in the nasal cavity is in the range of 0–900 ppb [11], while it is in the range of 20–100 ppb in the oral cavity. In addition, the sensor was characterized with ammonia ($NH_3$) gas, which is present in the oral cavity in the range of 0–450 ppb [12].

## 2. Sensor Design and Measurement Method

The sensor design is based on the principle of amperometric detection of $H_2S$ gas. The working, reference, and counter electrodes, together with the electrolyte, constitute the basic elements of the sensor. Particularly in the present design, the working electrode consists of a nanostructured Nafion™ (Chemours, Wilmington, DE, USA) coating that in turn is leveraged through a microporous high aspect ratio structure. The interaction between this large-area working electrode, the gas, and the electrolyte (5% $H_2SO_4$) under electrical bias leads to the oxidation of $H_2S$ gas at the surface of the electrode, causing a current flow between the working and the counter electrodes. The working electrode current is then measured using a potentiostat, maintaining a constant voltage of +1.1 V with respect to the reference electrode. Figure 1a shows the schematic cross section of the sensor design, and Figure 1b shows the photograph of a bare die of the sensor with a dimension of 10 mm × 10 mm × 1 mm. The design, fabrication, and assembly of the sensor have been described in our previous work [13].

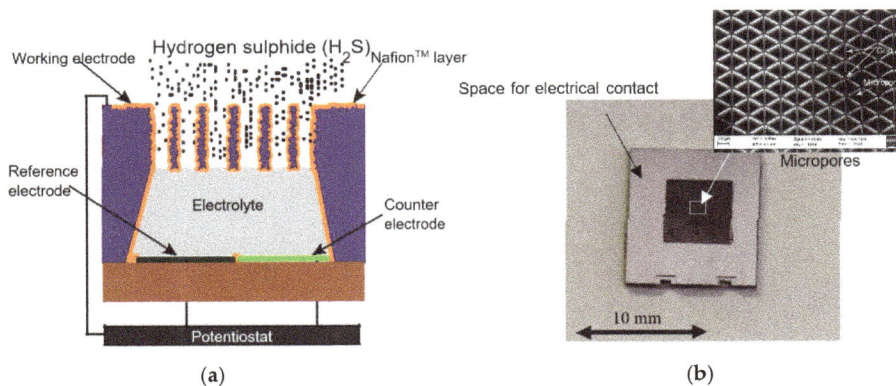

(a)                                                    (b)

**Figure 1.** (a) Schematic cross section of the sensor design indicating the high aspect ratio micropores with a nanostructured porous Nafion™ layer. (b) Photograph of the amperometric sensor with the porous structure in the middle (dark area). Inset: SEM image shows the microporous grid structure of the working electrode. The working electrode of the sensor is fabricated by deep reactive ion etching and platinum atomic layer deposition of a silicon on insulator (SOI) wafer, and the counter and reference electrodes are fabricated on a glass wafer, which is then assembled together by anodic bonding [13]. The sensor has a footprint area of $10 \times 10$ mm$^2$ and a thickness of approximately 1 mm.

In order to test the sensor for different gases and gas concentrations, a measurement set-up was built as illustrated in Figure 2. In this set-up, a 10 ppm $H_2S$ in $N_2$ gas (AGA gas AB, Lidingö, Sweden) was mixed with a pure $N_2$ gas (99.95% pure, AGA gas AB, Lidingö, Sweden) to obtain the desired concentration. A scrubber (Dräger, type 1140, Lidingö, Sweden) was used to remove potential residues in the $N_2$ gas. To measure the selectivity of the sensor to interfering gases, 200 ppb NO in $N_2$ gas (AGA gas AB, Lidingö, Sweden) and 45 ppm of $NH_3$ in $N_2$ gas (AGA gas AB, Lidingö, Sweden) were used. In order to humidify the gas mixture, a custom-made humidifier consisting of a syringe with moistened paper was used. A mechanical sealing module was further used to reduce evaporation of the electrolyte. Further details on the measurement set-up used for sensor characterization can be found in our previous work [13].

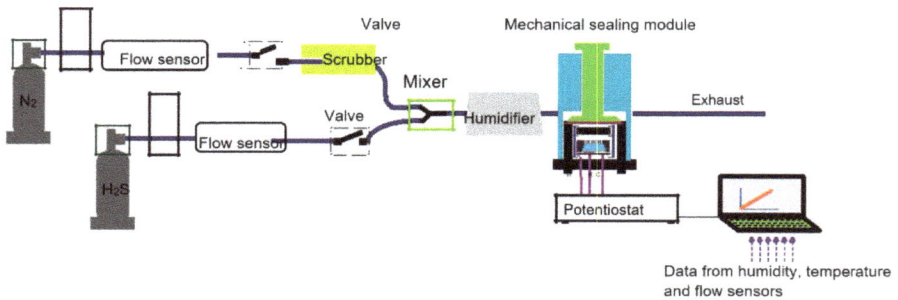

**Figure 2.** Schematic illustration of the measurement set-up used for characterization of the $H_2S$ amperometric sensor. Data from the flow sensors, the temperature sensor, and the humidity sensor were accessed using a LabVIEW ™ (National Instruments, Austin, TX, USA) program.

## 3. Results and Discussion

The sensor was tested for its $H_2S$ gas sensitivity, selectivity to NO and $NH_3$, and response time. A gas flow of 550 mL/min and 50% relative humidity (RH) was maintained for all measurements.

### 3.1. Sensitivity

In order to determine the sensitivity of the sensor, the $H_2S$ concentration was varied in five steps of 75, 150, 250, 500, and 820 ppb. Three such variations were carried out using the measurement set-up. The output current from each concentration was determined by calculating the difference between the working electrode current at $t_{90}$ (cf. Figure 5) and the working electrode current at zero $H_2S$ concentration. The output currents and the linear fitting for five different $H_2S$ gas concentrations are plotted in Figure 3. Based on the linear fit, the maximum sensitivity of the sensor was calculated to be 0.65 nA/ppb. The sensor detects $H_2S$ gas in the lower limit of 75 ppb and a higher limit of 820 ppb and is within the range required for monitoring the oral breath. The lower limit concentration of 75 ppb was measured without being affected by the noise in the system.

The sensitivity graph of commercial $H_2S$ sensors as compared to the designed integrated sensor is shown in Figure 4. In order to maintain a fair comparison, the area of the working electrode is normalized to the sensitivity, i.e., 0.65 nA/ppb is obtained for a footprint area of 25 mm$^2$ arriving at a normalized sensitivity of 2700 nA/ppm/cm$^2$. The integrated $H_2S$ sensor has a measured area normalized sensitivity that is approximately 2.5 times more than the sensitivities of commercial sensors. This shows a potential for the integrated sensor to be fabricated with a smaller footprint area with a reduced sensitivity in order to realize a smaller sensor, leading to a miniaturized handheld instrument.

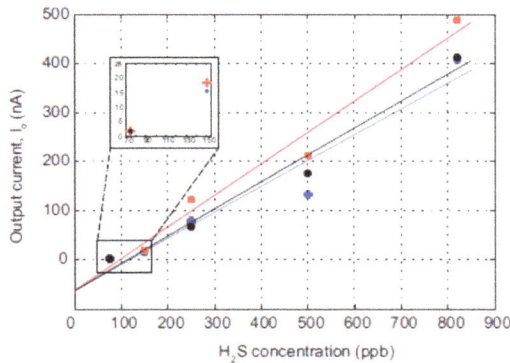

**Figure 3.** Output current $I_o$ as a function of $H_2S$ concentration. A linear fit to each concentration variation gives the slope, which is then related to the sensitivity of the sensor.

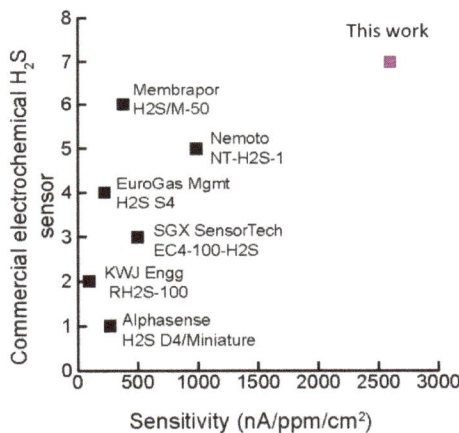

**Figure 4.** Comparison of area-normalized $H_2S$ sensitivity ranges for several commercial amperometric sensors. The sensitivity of the integrated sensor is approximately 2.5 times higher than the sensitivity of commercial sensors.

*3.2. Selectivity to NO and NH₃*

To obtain the selectivity to NO gas, the output current was measured at 200 ppb NO gas concentration, and the NO sensitivity was calculated to be approximately 0.04 nA/ppb, which is in agreement with our earlier fabricated prototype [13,14]. The selectivity can be defined as the ratio of $H_2S$ sensitivity to NO sensitivity and is calculated to be approximately 16. NO concentration of approximately 900 ppb is commonly found in the nasal cavity that can affect the oral breath [11]. However, a concentration of 900 ppb of NO results in an equivalent $H_2S$ concentration of 55 ppb, which is within a normal halitosis range. Therefore, the NO contamination from the nasal cavity has negligible effect on the $H_2S$ concentration from the oral breath. The selectivity of the sensor to NO could be further increased by reducing the nasal NO contamination from the oral cavity either by breath maneuver or by clamping the nostrils while measuring $H_2S$ concentration from the oral cavity. Other sources of NO release, such as lungs and oral cavity, can be neglected due to a low NO concentration (20–100 ppb), which is likely to have a minimal interference with $H_2S$ detection.

The output current of the sensor to 45 ppm $NH_3$ gas concentration was found to be below the detection limit. Therefore, the sensor will not be sensitive to $NH_3$ gas present in the oral cavity.

## 3.3. Response Time

The response time of the sensor to a 250 ppb $H_2S$ concentration was estimated by measuring the rise time ($t_{90}$) of the sensor, i.e., the time required to reach 90% of the maximum output current. The response time of the sensor was measured to be 21 s, as shown in Figure 5. A thorough investigation of the method of oral breath sampling to capture the $H_2S$ gas concentration at a constant rate that could be correlated to the sensor response time has not yet been reported. However, Tangerman et al. collected a sample of oral breath by breathing into a syringe for 5–10 s. The sample was then used for $H_2S$ concentration detection [15]. The recommended procedure for Halimeter instrument includes an initial three-minute period during which the patient breathes through the nose with lips sealed. A pipe attached to the instrument is then inserted into the partially opened mouth, and a pump withdraws gas from the oral cavity for concentration measurement [16,17]. It is conceivable that an oral breathing in the range of 20–30 s could be considered for $H_2S$ concentration measurement.

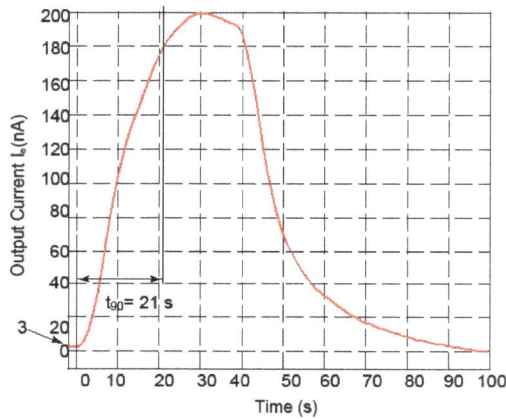

**Figure 5.** Output current response to 250 ppb step of $H_2S$ gas concentration. A response time ($t_{90}$), i.e., the time required to reach 90% of the maximum output current, of 21 s was measured for the sensor.

In order to assess the response time of the integrated sensor, a comparison graph with commercial $H_2S$ sensors is shown in Figure 6. The response time ($t_{90}$) of 21 s is comparatively better than most of the currently available commercial sensors. In order to have a real-time measurement, the response time can be further decreased by optimization of the sensor design.

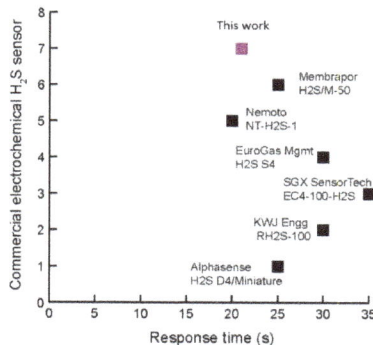

**Figure 6.** Response times of several commercial amperometric $H_2S$ sensors. The graph shows that the integrated sensor has a better response time compared to commercial sensors.

### 3.4. Working Electrode Current Drift

The response of the sensor to five different concentrations is shown in Figure 7. The graph shows a background current drift of 0.675 nA/min, which is equivalent to approximately 1 ppb/min of $H_2S$ concentration drift. Therefore, for a response time of 21 s, the drift component is approximately 0.3 ppb, which is negligibly small. Hence, an accurate measurement of $H_2S$ concentration can be deducted from the output current.

### 3.5. Halitosis Measurement Range

Figure 7 shows the working electrode current for five different steps of $H_2S$ gas concentration. The sensor can measure the $H_2S$ concentration in the region of normal, weak, and strong halitosis that is required for monitoring the oral health. Consequently, the sensor can be applied to detect the entire dynamic range of $H_2S$ concentration present in the oral breath. In addition, the sensor could be useful in dental clinics for follow-up measurement of the halitosis content before and after oral dental treatment.

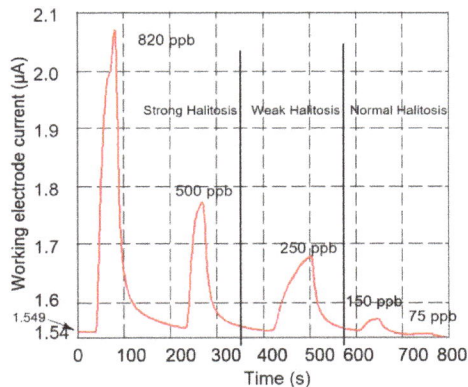

**Figure 7.** Working electrode current for five different $H_2S$ concentrations. The sensor is able to detect the entire dynamic range of $H_2S$ concentration that could be correlated to the concentration present in the oral breath. The background current of the sensor could be due to the interference from humidity or due to high conductivity of the electrolyte.

## 4. Conclusions

In this paper, an integrated amperometric sensor has been evaluated for the detection of hydrogen sulfide ($H_2S$) concentration present in the oral breath. The sensitivity of the sensor is measured to be 0.65nA/ppb with a response time of approximately 21 s, which is comparatively better than commercially available sensors. The sensor can be applied to measure bad breath where the concentration of $H_2S$ gas indicates a malodor in the breath. The entire range of $H_2S$ gas concentration present in the oral breath can be diagnosed by the sensor. For more complete measurements, detection of methyl mercaptan ($CH_4S$) and dimethyl sulfide ($CH_3)_2S$ concentration in combination with $H_2S$ concentration will give a more comprehensive representation of oral health. Overall, a prototype has been realized to measure $H_2S$ gas concentration that is relevant for bad breath monitoring.

**Author Contributions:** Conceptualization, H.K.G., G.S and N.R.; Methodology, H.K.G, N.R; Software, H.K.G; Validation, H.K.G, G.S. and N.R.; Formal Analysis, H.K.G, G.S. and N.R; Investigation, H.K.G; Resources, G.S. and N.R.; Data Curation, N.R.; Writing-Original Draft Preparation, H.K.G; Writing-Review & Editing, H.K.G., G.S. and N.R; Visualization, H.K.G; Supervision, G.S. and N.R.; Project Administration, N.R.; Funding Acquisition, G.S. and N.R.

**Funding:** This work was supported by the Swedish agency VINNOVA, the European Research Council (ERC), through Advanced Grant No: 267528.

**Acknowledgments:** H.K.G. would like to thank Dr. Santhosh Kumar, Associate Professor, Department of Periodontics, College of Dental Sciences, K.M.C Manipal, Manipal, INDIA, for valuable discussions on oral health and diseases.

**Conflicts of Interest:** The authors are inventors on a granted patent and multiple patent applications describing a miniaturized amperometric gas sensor.

## References

1. Yokoyama, S.; Ohnuki, M.; Shinada, K.; Ueno, M.; Wright, C.; Allan, F. Oral malodor and related factors in Japanese senior high school students. *J. Sch. Health* **2010**, *80*, 346–352. [CrossRef] [PubMed]

2. Apatzidou, A.D.; Bakirtzoglou, E.; Vouros, I.; Karagiannis, V.; Papa, A.; Konstantinidis, A. Association between oral malodour and periodontal disease-related parameters in the general population. *Acta Odontol. Scand.* **2013**, *71*, 189–195. [CrossRef] [PubMed]

3. Van den Broek, M.; Feenstra, L.; de Baat, C. A review of the current literature on aetiology and measurement methods of halitosis. *J. Dent.* **2007**, *35*, 627–635. [CrossRef] [PubMed]

4. Zürcher, M.; Laine, L.; Filippi, A. Diagnosis, prevalence, and treatment of halitosis. *Curr. Oral Health Rep.* **2014**, *1*, 279–285. [CrossRef]

5. The Halimeter®—Measure Bad Breath Scientifically. 2015. Available online: http://www.halimeter.com/the-halimeter-measure-bad-breath-scientifically/ (accessed on 8 January 2015).

6. Baharvand, M.; Maleki, Z.; Mohammadi, S.; Alavi, K.; Moghaddam, E. Assessment of oral malodor: A comparison of the organoleptic method with sulfide monitoring. *J. Contemp. Dent. Pract.* **2008**, *9*, 76–83. [PubMed]

7. Donaldson, A.C.; Riggio, M.; Rolph, H.; Bagg, J.; Hodge, P. Clinical examination of subjects with halitosis. *Oral Dis.* **2007**, *13*, 63–70. [CrossRef] [PubMed]

8. Chroma. Comparison between Oral Chroma and Halimeter. Available online: http://oralchroma.es/catalogos/OralChroma-vs-Halimeter-eng.pdf (accessed on 8 January 2015).

9. Schiavon, G.; Zotti, G.; Toniolo, R.; Bontempelli, G. Electrochemical detection of trace hydrogen sulfide in gaseous samples by porous silver electrodes supported on ion-exchange membranes (solid polymer electrolytes). *Anal. Chem.* **1995**, *67*, 318–323. [CrossRef]

10. Yang, X.; Zhang, Y.; Hao, X.; Song, Y.; Liang, X.; Liu, F.; Sun, P.; Gao, Y.; Yan, X.; Lu, G.; et al. Nafion-based amperometric $H_2S$ sensor using Pt-Rh/C sensing electrode. *Sens. Actuators B Chem.* **2018**, *273*, 635–641. [CrossRef]

11. Horvath, I.; Loukides, S.; Wodehouse, T.; Csiszer, E.; Cole, P.; Kharitonov, S. Comparison of exhaled and nasal nitric oxide and exhaled carbon monoxide levels in bronchiectatic patients with and without primary ciliary dyskinesia. *Thorax* **2003**, *58*, 68–72. [CrossRef] [PubMed]

12. Schmidt, F.M.; Vaittinen, O.; Metsälä, M.; Lehto, M.; Forsblom, C.; Groop, P. Ammonia in breath and emitted from skin. *J. Breath Res.* **2013**, *7*, 017109. [CrossRef] [PubMed]

13. Gatty, H.K.; Stemme, G.; Roxhed, N. A wafer-level liquid cavity integrated amperometric gas sensor with ppb-level nitric oxide gas sensitivity. *J. Micromech. Microeng.* **2015**, *25*, 105013. [CrossRef]

14. Gatty, H.K.; Leijonmarck, S.; Antelius, M.; Stemme, G.; Roxhed, N. An amperometric nitric oxide sensor with fast response and ppb-level concentration detection relevant to asthma monitoring. *Sens. Actuators B Chem.* **2015**, *209*, 639–644. [CrossRef]

15. Tangerman, A.; Winkel, E. The portable gas chromatograph OralChroma™: A method of choice to detect oral and extra-oral halitosis. *J. Breath Res.* **2008**, *2*, 017010. [CrossRef] [PubMed]

16. Furne, J.; Majerus, G.; Lenton, P.; Springfield, J.; Levitt, D.; Levitt, M. Comparison of volatile sulfur compound concentrations measured with a sulfide detector vs. gas chromatography. *J. Dent. Res.* **2002**, *81*, 140–143. [CrossRef] [PubMed]

17. Interscan Corp., Simplified RH17K Manual. Available online: http://www.halimeter.com/images/Simplified_RH17K_Manual.PDF (accessed on 8 January 2015).

*micromachines*

MDPI

*Article*

# A Set of Platforms with Combinatorial and High-Throughput Technique for Gas Sensing, from Material to Device and to System

**Zhenghao Mao [1], Jianchao Wang [2], Youjin Gong [1], Heng Yang [2] and Shunping Zhang [2,3,*]**

[1] Institute of Nuclear Physics and Chemistry, China Academy of Engineering Physics, Mianyang 621900, China; zh.mao@foxmail.com (Z.M.); gyjlyq@163.com (Y.G.)

[2] Nanomaterials and Smart Sensors Research Laboratory, Department of Materials Science and Engineering, Huazhong University of Science and Technology, Wuhan 430074, China; 15827258559@163.com (J.W.); yy9512812@163.com (H.Y.)

[3] Shenzhen Institute of Huazhong University of Science & Technology, Shenzhen 518000, China

\* Correspondence: pszhang@mail.hust.edu.cn; Tel.: +86-159-2711-4217

Received: 20 September 2018; Accepted: 12 November 2018; Published: 19 November 2018

**Abstract:** In a new E-nose development, the sensor array needs to be optimized to have enough sensitivity and selectivity for gas/odor classification in the application. The development process includes the preparation of gas sensitive materials, gas sensor fabrication, array optimization, sensor array package and E-nose system integration, which would take a long time to complete. A set of platforms including a gas sensing film parallel synthesis platform, high-throughput gas sensing unmanned testing platform and a handheld wireless E-nose system were presented in this paper to improve the efficiency of a new E-nose development. Inkjet printing was used to parallel synthesize sensor libraries (400 sensors can be prepared each time). For gas sensor selection and array optimization, a high-throughput unmanned testing platform was designed and fabricated for gas sensing measurements of more than 1000 materials synchronously. The structures of a handheld wireless E-nose system with low power were presented in detail. Using the proposed hardware platforms, a new E-nose development might only take one week.

**Keywords:** combinatorial and high-throughput technique; array optimization; electronic nose; efficiency

## 1. Introduction

The electronic nose (E-nose) is an analytical device that plays a constantly growing role as a general purpose detector of vapor chemicals in many applications such as the quality control of the food industry [1,2], environmental protection [3–5], public safety [6] and spaceflight applications [7], according to articles on the subject that have been published over the last fifteen years. The core component in E-noses is the gas sensor array, which is made up of several gas sensors with different gas sensing properties to improve selectivity to gas/odor [8,9]. For a specific E-nose application, the gas sensor array needs to be optimized to the appropriate size and components [10–12]. There are several essential processes in E-nose development: the preparation of gas sensitive materials, gas sensor fabrication, array optimization, sensor array package and E-nose system integration. All these processes have an impact on the properties of sensitivity, selectivity, stability, price and power of the E-nose system [13]. Aside from the 3S2P (Sensitivity, Selectivity, Stability, Price and Power) properties, another factor that also needs to be considered is "Efficiency." In other words, how long these processes would take to develop a new application, one month or one week?

For a new application, it takes a lot of time to find the optimized sensor array. The optimized sensor array could be selected from a sensor library that contains many gas sensors with different sensing properties [10–12]. The gas sensitive materials of these sensors could be synthesized through many methods, for example, a metal oxide semiconductor could be used as a gas sensitive material such as $SnO_2$, $ZnO$, $WO_3$ and $In_2O_3$ [14–18]. However, it would take a long time to fabricate the sensor library. The gas sensing properties of these sensors need to be measured according to the gas/odor to be classified in the E-nose application, which also takes a long time. After optimizing the sensor array, the E-nose system needs to be integrated based on the optimized sensor array and application. The E-nose system's measuring range of sensor signals should be large enough to optimize the array. For gases/odor identification, the knowledge database of the E-nose system needs to be comprehensive to the optimized array and the gases/odor in the application. Accordingly, to improve the "Efficiency" in E-nose development, three processes should be considered: material parallel synthesis, high throughput screening and general utility of the E-nose system.

Parallel synthesis and high throughput screening (called combinatorial and high-throughput technique) of gas sensitive materials could accelerate the selection of gas sensitive materials with good sensitivity, selectivity and stability. It could also be used in sensor array optimization. The difference between using the combinatorial method in materials selection and array optimization is the parallel synthesizing of gas sensitive materials or gas sensing films of sensor devices. For example, the Simon group used sol-gel and polyol methods to parallel synthesize gas sensitive materials [19–22]. Sol-gel and polyol methods are material synthesis methods and may not be suitable for gas sensing film synthesis in the batch production of sensor devices. The traditional thick film forming method of gas sensor fabrication is screen printing, however, it takes a long time to prepare the paste for screen printing and only one sensing film can be printed at a time.

In this paper, the inkjet printing technique and a high throughput gas sensing unmanned testing platform for the signal measurement of the sensor library were presented. After gas sensing films were parallel synthesized on the substrate, the sensor array (substrate with sensing films) needs to be packaged into a gas sensor for the acquisition of further performance. A simple sensor array package structure with low power and low price was also detailed in this paper. A handheld wireless E-nose system with the ability to measure 100 $\Omega$ to 1 $G\Omega$ sensor resistance was also described. With the above platforms, the sensor array could be easily fabricated, optimized, packaged and assembled into an E-nose system for a new E-nose application within one week.

## 2. Combinatorial and High-Throughput Technique for Screening Gas Sensor

### 2.1. Parallel Synthesis of Sensor Library

The application of a combinatorial and high-throughput technique in materials research promises significant acceleration, especially in the area of material and parameter optimization as well as in the discovery of new materials. The first and significant step of the combinatorial and high-throughput technique is the parallel synthesis of different gas sensitive materials. The combinatorial and high-throughput technique could also be used in gas sensor selection and array optimization. The first step may not be the parallel synthesis of different materials but the parallel synthesis of different gas sensing films, which could be used in the batch production of sensor devices. This is because the optimized array is a sensor device for batch production and E-nose application. Many parallel synthesis methods of gas sensitive materials such as sol-gel and polyol methods could not be directly used to parallel synthesize gas sensing films in gas sensors.

For the parallel synthesis of gas sensing films, a platform named the gas sensing film parallel synthesis platform (shown in Figure 1a) was designed and manufactured. The platform consisted of two main parts: the premix module and the transfer printing module. The premix module consisted of a raw material cavity, peristaltic pump array, droplet needle array, premixed chamber, blender array and so forth. The transfer printing module was composed of the transfer printing needle and deflection angle camera. The gas sensing film parallel synthesis platform is suitable for metal oxide gas sensing materials. Here, the experiment of precious metal ions modified $SnO_2$ is shown as an example to introduce the platform. As can be seen from Figure 2a, the SnO$_2$ peaks fit the rutile structure and no secondary phase was detected. All peaks indexed well to the SnO$_2$ JCPDS (Joint Committee on Poder Diffraction Standards) No. 88-0287. The high resolution transmission electron microscopy (HR-TEM; Tecnai G$^2$ F30, FEI Company, Eindhoven, Netherlands) image of pure SnO$_2$ is exhibited in Figure 2b where the particle size is approximately 15 nm. After preparing the gas sensitive material, jettable inks were made by dispersing SnO$_2$ into deionized water with ball milling and were collected in the raw material cavity. Then, premixed solutions with different components conserved in the premixed chamber were obtained by mixing SnO$_2$ inks and various additives (precious metal ionic solution) with the peristaltic pump and droplet needle array. To ensure the homogeneity of the premixed solutions, an array blender was used. A deflection angle camera was used to calibrate the deflection of the 8-matrices material substrate shown in Figure 3. The transfer printing needle was utilized to deposit the gas sensing films by transferring the premixed solutions to the 8-matrices material substrate. Finally, the 8-matrices material substrate was sintered for 2 h at 350 °C and 2 h at 550 °C and the gas sensing films with different components were parallel synthesized (see Figure 1b). Figure 1c–f show the scanning electron microscopy (SEM; Nova NanoSEM 450, FEI Company, Eindhoven, Netherlands) images with different magnifications of one modified SnO$_2$ gas sensing film. There were no macroscopic and microscopic cracks. The microscopic morphology of the tin dioxide was preserved. The performance indicators of the platform were as follows. Six kinds of jettable ink could be parallel mixed into 400 premixed solutions with different components. It took two minutes to prepare a single premixed solution and three minutes to transfer print a gas sensing film.

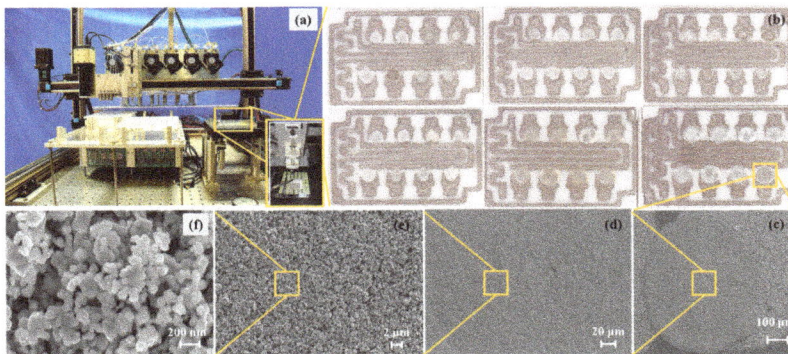

**Figure 1.** The parallel synthesis of gas sensing films. (**a**) The gas sensing film parallel synthesis platform. (**b**) The 8-matrices material substrate with sensing films. (**c–f**) Different magnifications of one gas sensing film.

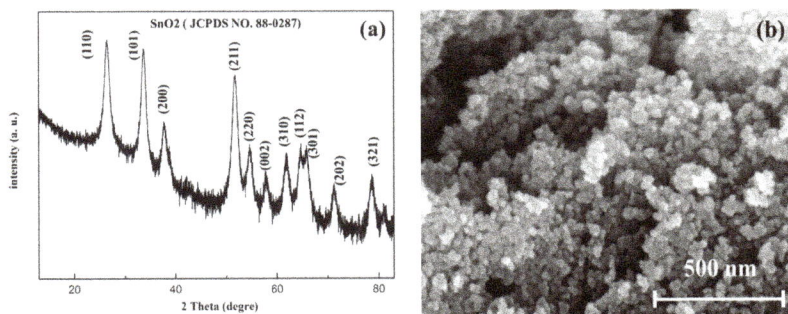

**Figure 2.** The X-ray diffraction (XRD) and scanning electron microscopy (SEM) images of the raw material. (**a**) X-ray diffraction pattern of SnO$_2$ nanoparticles. (**b**) SEM image of SnO$_2$ nanoparticles.

**Figure 3.** The structure of the 8-matrices material substrate.

## 2.2. High-Throughput Screening of Sensor Library

The next step for sensor selection and array optimization with the combinatorial and high-throughput technique is high throughput screening. A high throughput gas sensing unmanned testing platform (see Figure 4a) was designed.

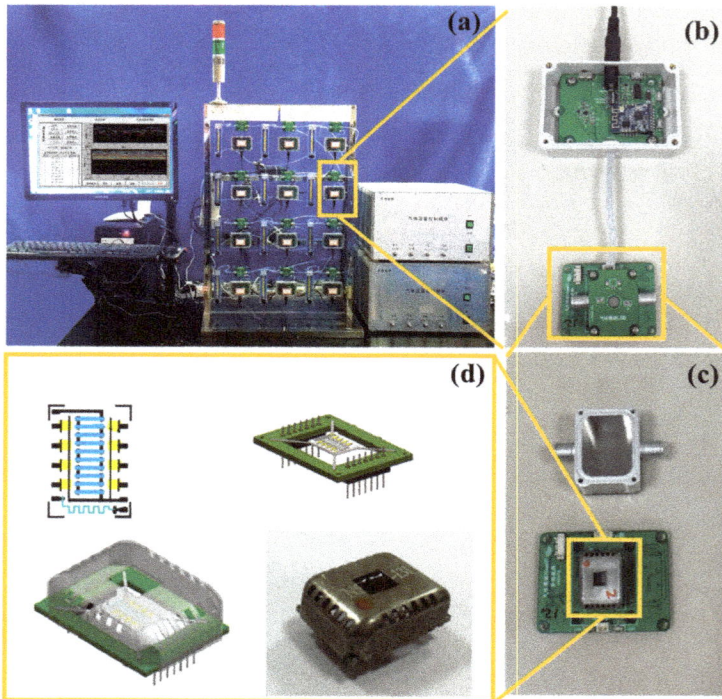

**Figure 4.** The structure of high-throughput gas sensing unmanned testing platform. (**a**) Testing platform; (**b**) Sensor module; (**c**) Testing chamber and sensor device; (**d**) The structure of the sensor device.

The testing network consisted of sensor modules (see Figure 4b) and an automatic gas mixing device. The sensor module was in charge of gas sensing, temperature modulation, temperature and light modulation and signal acquisition. Each module can be connected with a PC through a Wi-Fi communication module respectively or as a network. For resistance testing, the schematic diagram of the sensor module resistance test was as described in our previous work [23]. After depositing the gas sensing films and sintering, the 8-matrices material substrate was packaged into a low-power gas sensor device (see Figure 4d). For resistance measuring, the sensor device with a sensor module was connected (see Figure 4c). The testing environment in the chamber of every sensor module could be regulated by the automatic gas mixing device, which could mix the target gas with the carrier gas to a specific concentration through four parallel pressure sensors, mass flow controllers and flux valves. The mixed testing gas was equally distributed into 12 pathways through rotameters. A computer cooperating with software was used to operate the sensor modules and automatic gas mixing device. The automatic signal control, acquisition and preservation of each module and gas mixing device enabled unattended testing.

The performance indicators of the platform were as follows. The measurement range of resistance was 100 $\Omega$–1 G$\Omega$ with an error less than 5% [24–27]. The resistance acquisition rate of each channel was more than 2 Hz. According to the local area TP/IP network, the number of the sensor modules could be expanded to 254. That is to say, 2032 (8 $\times$ 254) gas sensing films could be parallel characterized. The temperature control range of the sensor module was between room temperature and 550 °C with an error less than 0.5 °C.

A high-throughput screening example of a sensor library using the platform is shown in Figure 5. The basic material of the sensor library was nano-$SnO_2$. The additive materials are shown in Table 1. As can be seen, seven kinds of precious metal ions were used for surface modification. The proportion of precious metal ions added is shown in Table 2. There were fifty-five sensing films with modified precious metal ions and one pure $SnO_2$ sensing film. The responses of the sensor library to CO, ethylene, benzene, formaldehyde, acetone and ethanol are shown in Figure 5. The concentration of all tested gases was 100 ppm. The tests were conducted at three constant temperature, 350 °C, 250 °C and 150 °C, respectively. According to the results, 350 °C was the optimum working temperature of all sensors. The response of all sensors at 350 °C is shown in Figure 5.

**Table 1.** The source of additives applied in the surface modification.

| Number | Element | Source | Number | Element | Source |
|--------|---------|--------|--------|---------|--------|
| 1 | Null | Null | 5 | Pd | $PdCl_2$ |
| 2 | Pt | $H_2PtCl·6H_2O$ | 6 | Ir | $IrCl_4$ |
| 3 | Rh | $RhCl_3$ | 7 | Au | $AuCl_3·HCl$ |
| 4 | Ru | $RuCl_3·xH_2O$ | 8 | Ag | $AgNO_3$ |

The results showed that the response to the tested gases of basic materials was improved evidently by the modified precious metal ions. The responses of pure $SnO_2$ to 100 ppm to the six tested gases (CO, ethylene, benzene, formaldehyde, ethanol and acetone) were 3.0, 4.6, 14.1, 29.3, 50.0 and 19.6, respectively, while the maximum responses to the six tested gases were 20.5, 21.3, 54.1, 1452.3, 2861.5 and 2973.5, obtained by $SnO_2$ + 0.05 mol% Ir, $SnO_2$ + 0.2 mol% Ir, $SnO_2$ + 0.05 mol% Ir, $SnO_2$ + 0.5 mol% Ir, $SnO_2$ + 0.5 mol% Ir and $SnO_2$ + 0.5 mol% Ir, respectively. It can be concluded that the precious metal ion Ir had the largest improvement in the gas sensing performance of tin oxide. Ir modified $SnO_2$ had a modest selectivity when sensing the six tested gases. This is because of the broad-spectrum response of the metal oxide gas sensitive materials. However, it could not distinguish the six tested gases based on the result shown in Figure 5.

**Table 2.** The proportion of added elements.

| Number | 1 | 2 | 3 | 4 | 5 | 6 | 7 | 8 |
|--------|---|---|---|---|---|---|---|---|
| Pt (mol%) | 0 | 0.05 | 0.09 | 0.13 | 0.2 | 0.3 | 0.4 | 0.5 |
| Rh, Ru, Pd, Ir, Au, Ag (mol%) | 0.05 | 0.09 | 0.13 | 0.2 | 0.3 | 0.4 | 0.5 | 0.6 |

For gas identification, eight materials were selected as the optimized sensor array. The standard of material selection was as follows. First, 10 materials with a higher response of each tested gas were preliminarily selected. The preliminary selected materials were sorted according to the number of repetitions. Finally, the top eight materials were selected as the optimized materials to form array optimization. The selected eight materials were $SnO_2 + 0.3$ Pt mol%, $SnO_2 + 0.3$ Rh mol%, $SnO_2 + 0.4$ Ru mol%, $SnO_2 + 0.3$ Pd mol%, $SnO_2 + 0.05$ Ir mol%, $SnO_2 + 0.5$ Ir mol%, $SnO_2 + 0.5$ Au mol% and $SnO_2 + 0.5$ Ag mol%, respectively. To prove the accuracy of the gas sensors, the R-t (resistance-time) of one gas sensor ($SnO_2 + 0.5$ Ir mol%) to the six tested gases is shown in Figure 6. The selectivity of the eight selected materials to 30 ppm tested gases is shown in Figure 7.

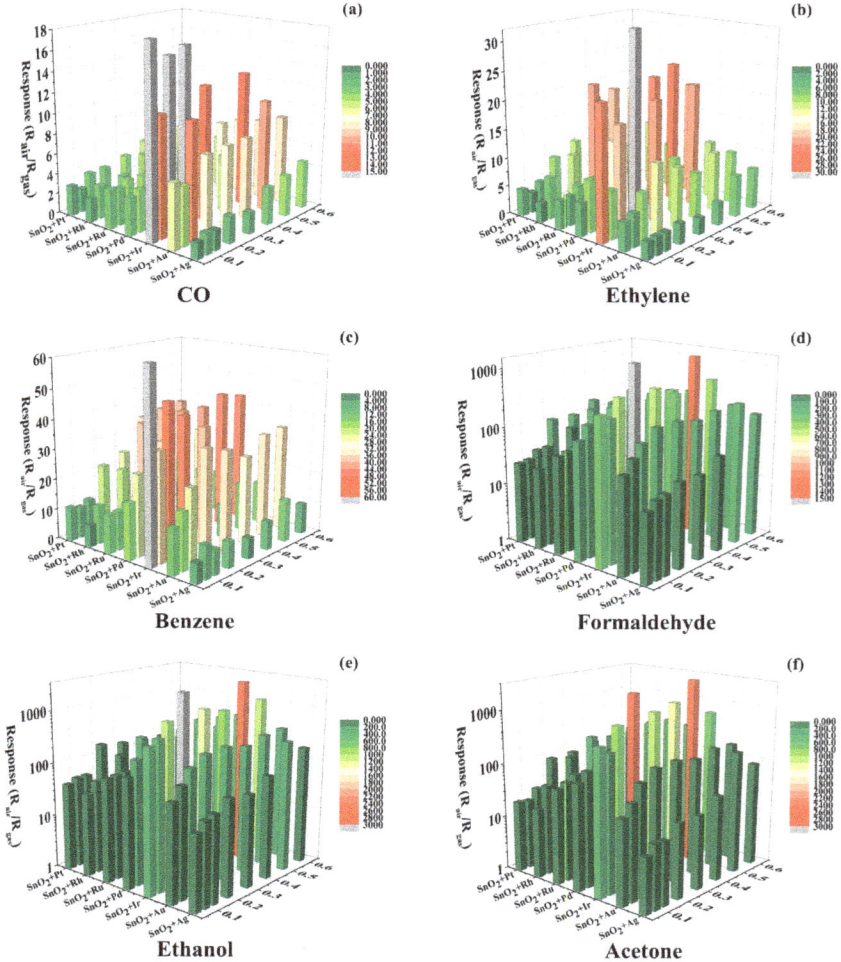

**Figure 5.** The high throughput resistance measurement (HT-RM) results of a sensor library based on nano-$SnO_2$ to 100 ppm of the six tested gases at 350 °C. (**a**) CO; (**b**) Ethylene; (**c**) Benzene; (**d**) Formaldehyde; (**e**) Ethanol; (**f**) Acetone.

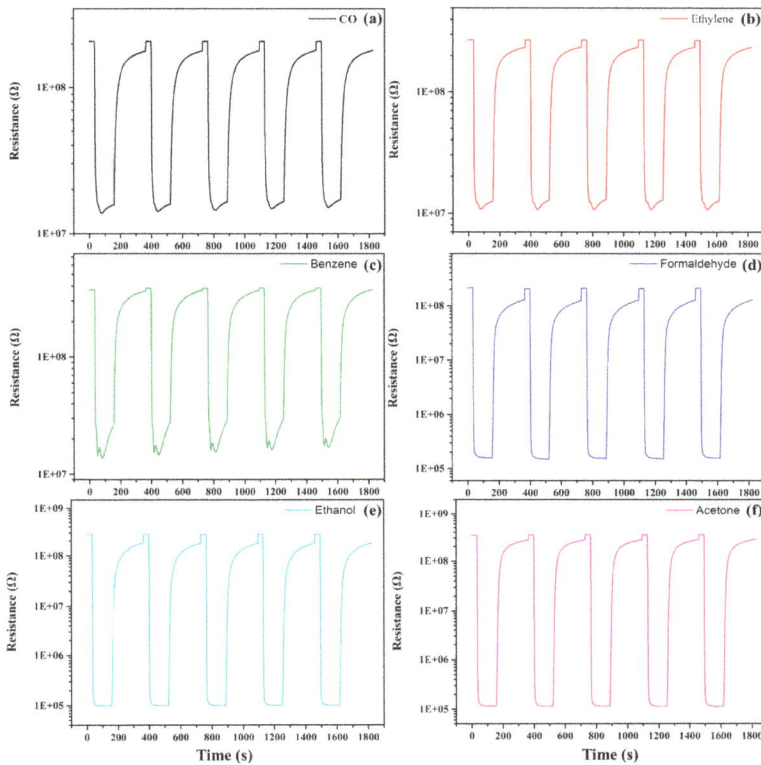

**Figure 6.** The Resistance–Time (R–T) curves of SnO$_2$ + 0.5 Ir mol% to 100 ppm of the six gases at 350 °C. (**a**) CO; (**b**) Ethylene; (**c**) Benzene; (**d**) Formaldehyde; (**e**) Ethanol; (**f**) Acetone.

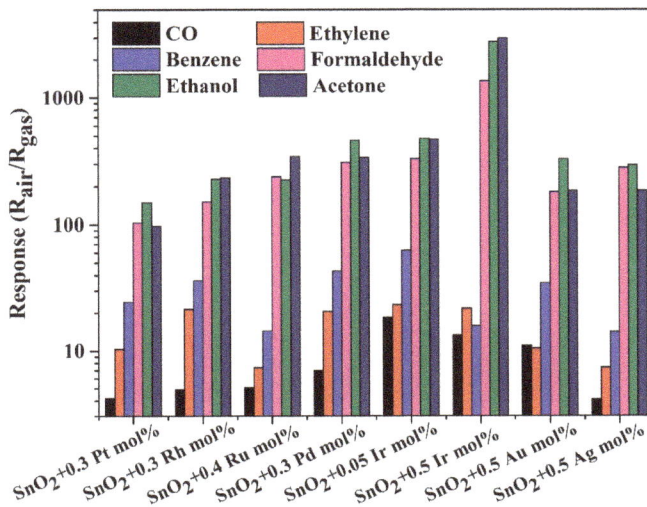

**Figure 7.** The selectivity of the eight selected materials to 100 ppm of the tested gases.

## 3. Details of Handheld Wireless E-Nose System

### 3.1. Handheld Wireless E-Nose System with Selected Materials

A handheld wireless E-nose system was designed to interface with the selected sensor array. The E-nose system mainly consisted of three parts: a pumping module, communication and power supply module and display record module, as shown in Figure 8. As can be seen, the selected materials were packed directly into the sensor device, followed by the combining sensor device and pumping module. The pumping module is in charge of the gas sensing, temperature modulation, temperature and light modulation and the environment in the testing chamber. The resistance measuring range was 100 Ω to 1 GΩ with an error less than 5%. The communication and power supply module was the power source and responsible for data acquisition and transmission and could be connected with a PC through the Wi-Fi communication module. For the display record and module, it interacted with a human through keys and the Liquid Crystal Display (LCD). The interface module also had a wireless interface with a PC for database update.

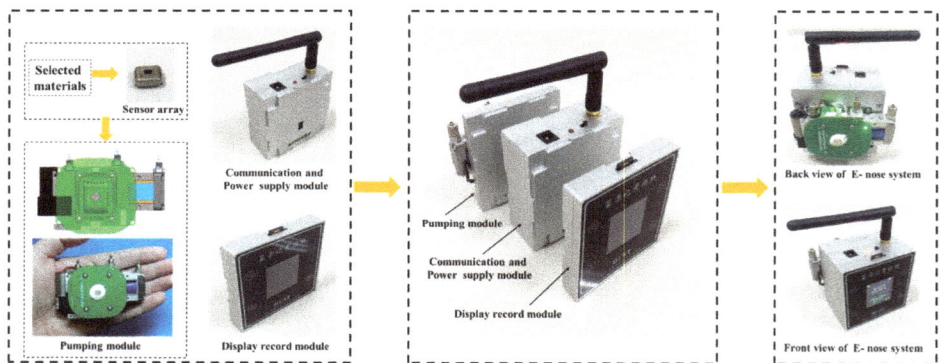

**Figure 8.** The structure of the handheld wireless E-nose system.

### 3.2. Application in Discrimination of Gases

The optimized sensor array (contained the selected eight materials) and handheld wireless E-nose system were combined for gas identification. Six gases with three concentrations of each gas were tested at 350 °C (optimum working temperature) by the handheld wireless E-nose system. Each sample was tested ten times. In total, $6 \times 3 \times 10 = 180$ samples were tested. All samples were tested in a random sequence and the experiment was conducted in one week. Then, the response of each sample was extracted as the original features. The resistance response of a certain sensor to the tested gas at optimum working temperature represented the information of the reactions between them. To eliminate the testing error, two tests with the maximum and minimum fluctuations of the feature values in each sample test were excluded. A total of $28 \times 8$ dimensions original feature space were extracted for each sample. Then, the dimensionality of the original was reduced by principal component analysis (PCA) and the tested gases were discriminated by Fisher discriminant analysis (FDA).

As can be seen from Figure 9, the samples of every tested gas were distributed in relatively different areas. Due to the concentration interference, part of the areas of acetone, formaldehyde, alcohol, benzene and ethylene overlapped and the distribution of acetone, formaldehyde and alcohol were not concentrated, so it could be not discriminated correctly. Furthermore, FDA was used as pattern recognition method. FDA has been widely used as a method of pattern recognition [24,25,28] and mainly constructs high-dimensional historical data into a low-dimensional principal component space. When discriminating, the data acquired in real time are projected onto the principal component space and transformed into new data for pattern recognition. The FDA can be divided into three steps. First, a category pattern vector that can express the data category is obtained. The feature vector that is the most important and sensitive feature parameter the data category is then extracted. Finally, the discriminant function composed of the feature vector is used for pattern recognition. Therefore, functions 1 and 2 are the discriminant function. In order to avoid over-fitting, K-fold cross-validation [28–30] was used to split the pretreated data into training and validation sets recurrently. Due to the large dataset, three-fold cross-validation was used. That is to say, the tested samples of each gas were divided into three subsamples. One subsample was used as the validation data for discrimination and the remaining two were used as training data. The FDA was repeated three times (the folds) to ensure that each of the three subsamples were used exactly once as the validation data. The two-dimensional plots of FDA are shown in Figure 10. As can be seen from Figure 10, the gas species was distributed independently. The six tested gases could be discriminated 100% correctly. The discrimination rates are shown in Table 3. The results showed that the handheld wireless E-nose system was reliable and repeatable.

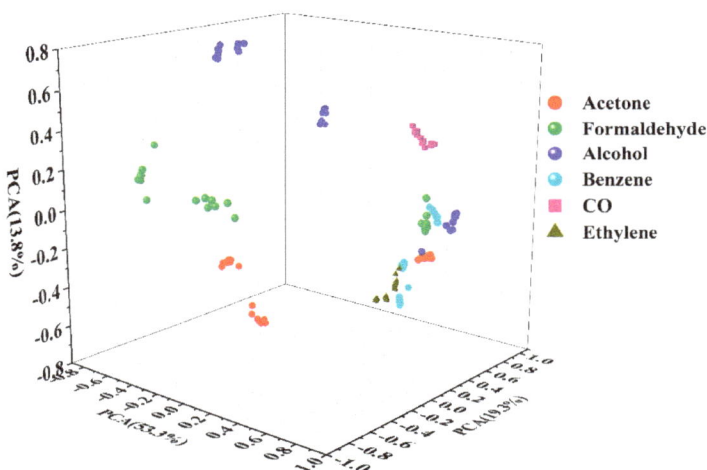

**Figure 9.** The principal component analysis (PCA) score plots of the sensing films.

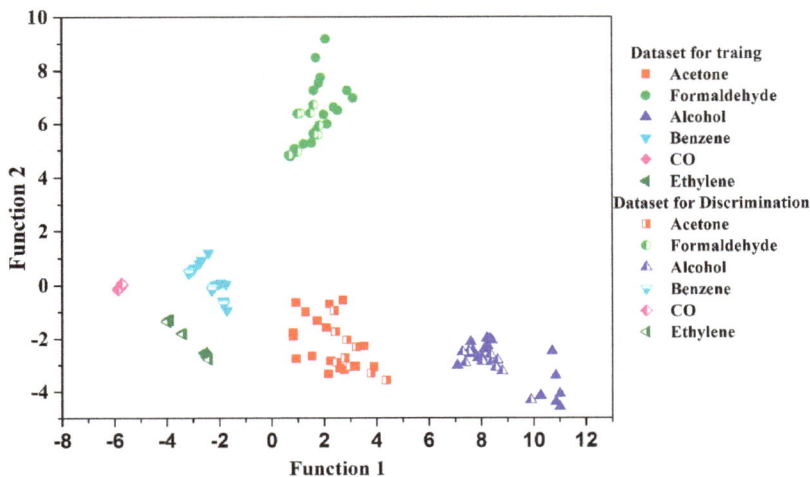

**Figure 10.** The Fisher discriminant analysis (FDA) results of all samples.

**Table 3.** The classification rates of FDA (Groups 1–6 represent CO, ethylene, benzene, ethanol, acetone and formaldehyde, respectively).

| Data of Each Gas for Training | Data of Each Gas for Discrimination | Classification Rates (%) | | | | | | |
|---|---|---|---|---|---|---|---|---|
| | | Group 1 | Group 2 | Group 3 | Group 4 | Group 5 | Group 6 | All Samples |
| 9–16, 17–24 | 1–8 | 100 | 100 | 100 | 100 | 100 | 100 | 100 |
| 1–8, 17–24 | 9–16 | 100 | 100 | 100 | 100 | 100 | 100 | 100 |
| 1–8, 9–16 | 17–24 | 100 | 100 | 100 | 100 | 100 | 100 | 100 |
| The Average Classification Rates | | 100 | 100 | 100 | 100 | 100 | 100 | 100 |

## 4. Conclusions

In this study, a set of hardware platforms to improve the efficiency of new E-nose development was presented including a gas sensing film parallel synthesis platform, high-throughput gas sensing unmanned testing platform and a handheld E-nose system. Inkjet printing was used as a parallel synthesis method to parallel synthesize the sensor libraries for gas sensor selection and array optimization. A high-throughput gas sensing unmanned testing platform was designed to measure the sensor library resistances within the range 100 $\Omega$ to 1 G$\Omega$. After array optimization among the sensors in sensor libraries, the selected sensors could be easily packaged into an eight-sensor array sensor device with low power under the working temperature from room temperature to 550 °C with an error less than 0.5 °C. A handheld wireless E-nose system was designed to interface with one sensor device.

Fifty-five SnO$_2$ sensing films with surface modification and an intrinsic SnO$_2$ film were parallel deposited by the gas sensing film parallel synthesis platform. The gas sensing performance of the sensor library was obtained by the gas sensing unmanned testing platform. With the above two platforms, the screening efficiency of a suitable sensor selected from the sensor libraries was greatly improved. A portable E-nose system assembled with a suitable sensor was used for gas identification. With this set of hardware platforms, the future directions of this work include the screening efficiency of high-performance gas sensitive materials. Sensor arrays of various sensing performances will be available and can be used in many fields (food industry, environment, public safety).

**Author Contributions:** Conceptualization, Data curation, Z.M.; Formal analysis, Z.M.; Writing—original draft preparation, Z.M.; Methodology, J.W.; Validation, J.W. and Y.G.; Investigation, H.Y.; Software, S.Z.; Resources, S.Z.; Supervision, S.Z.; Project administration, S.Z.; Funding acquisition, S.Z.

**Funding:** This research was funded by the Nature Science Foundation of China (NO. 61771207, NO. 61604142), the Natural Science Foundation of Hubei Province (ZRMS2017000373) and the Shenzhen Research Council (JCYJ20160506170101603).

**Acknowledgments:** The authors are also grateful to the Analytical and Testing Center of the Huazhong University of Science and Technology.

**Conflicts of Interest:** The authors declare no conflict of interest.

## References

1. Loutfi, A.; Coradeschi, S.; Mani, G.K.; Shankar, P.; Rayappan, J.B.B. Electronic noses for food quality: A review. *J. Food Eng.* **2015**, *144*, 103–111. [CrossRef]
2. Feng, T.; Zhuang, H.; Ye, R.; Jin, Z.; Xu, X.; Xie, Z. Analysis of volatile compounds of Mesona Blumes, gum/rice extrudates via GC–MS and electronic nose. *Sens. Actuators B Chem.* **2011**, *160*, 964–973. [CrossRef]
3. Romain, A.; Nicolas, J. Long Term Stability of Metal Oxide-Based Gas Sensors for E-nose Environmental Applications: An overview. *Sens. Actuators B Chem.* **2010**, *146*, 502–506. [CrossRef]
4. Delgado-Rodríguez, M.; Ruiz-Montoya, M.; Giraldez, I.; López, R.; Madejón, E.; Díaz, M.J. Use of electronic nose and GC-MS in detection and monitoring some VOC. *Atmos. Environ.* **2012**, *51*, 278–285. [CrossRef]
5. Tonacci, A.; Corda, D.; Tartarisco, G.; Pioggia, G.; Domenici, C. A Smart Sensor System for Detecting Hydrocarbon Volatile Organic Compounds in Sea Water. *CLEAN Soil Air Water* **2015**, *43*, 147–152. [CrossRef]
6. Norman, A.; Stam, F.; Morrissey, A.; Hirschfelder, M.; Enderlein, D. Packaging effects of a novel explosion-proof gas sensor. *Sens. Actuators B Chem.* **2003**, *95*, 287–290. [CrossRef]

7. Young, R.C.; Buttner, W.J.; Linnell, B.R.; Ramesham, R. Electronic nose for space program applications. *Sens. Actuators B Chem.* **2003**, *93*, 7–16. [CrossRef]
8. Stetter, J.R.; Penrose, W.R. Understanding chemical sensors and chemical sensor arrays (electronic noses): Past, present, and future. *Sens. Update* **2002**, *10*, 189–229. [CrossRef]
9. Wilson, D.M.; Garrod, S.; Hoyt, S.; McKennoch, S.; Booksh, K.S. Array optimization and preprocessing techniques for chemical sensing Microsystems. *Sens. Update* **2002**, *10*, 77–106. [CrossRef]
10. Zhang, T.; Sun, Q.; Yang, L.; Yang, L.; Wang, J. Vigor detection of sweet corn seeds by optimal sensor array based on electronic nose. *Trans. Chin. Soc. Agric. Eng.* **2017**, *33*, 275–281.
11. Chaudry, A.N.; Hawkin, T.M.; Travers, P.J. A method for selecting an optimum sensor array. *Sens. Actuators B Chem.* **2000**, *69*, 236–242. [CrossRef]
12. Zhang, S.; Xie, C.; Zeng, D.; Li, H.; Liu, Y.; Cai, S. A sensor array optimization method for electronic noses with sub-arrays. *Sens. Actuators B Chem.* **2009**, *142*, 243–252. [CrossRef]
13. Korotcenkov, G. Metal oxides for solid-state gas sensors: What determines our choice? *Mater. Sci. Eng. B* **2007**, *139*, 1–23. [CrossRef]
14. Ding, X.; Zeng, D.; Zhang, S.; Xie, C. C-doped $WO_3$ microtubes assembled by nanoparticles with ultrahigh sensitivity to toluene at low operating temperature. *Sens. Actuators B Chem.* **2011**, *155*, 86–92. [CrossRef]
15. Lupan, O.; Ursaki, V.V.; Chai, G.; Chow, L.; Emelchenko, G.A.; Tiginyanu, I.M.; Gruzintsev, A.N.; Redkin, A.N. Selective hydrogen gas nanosensor using individual ZnO nanowire with fast response at room temperature. *Sens. Actuators B Chem.* **2010**, *144*, 56–66. [CrossRef]
16. Yu, X.; Zeng, W. Fabrication and gas-sensing performance of nanorod-assembled $SnO_2$, nanostructures. *J. Mater. Sci. Mater. Electron.* **2016**, *27*, 7448–7453. [CrossRef]
17. Urasinska-Wojcik, B.; Vincent, T.A.; Chowdhury, M.F.; Gardner, J.W. Ultrasensitive $WO_3$, gas sensors for $NO_2$, detection in air and low oxygen environment. *Sens. Actuators B Chem.* **2017**, *239*, 1051–1059. [CrossRef]
18. Anand, K.; Kaur, J.; Singh, R.C.; Thangaraj, R. Preparation and characterization of Ag-doped $In_2O_3$ nanoparticles gas sensor. *Chem. Phys. Lett.* **2017**, *682*, 140–146. [CrossRef]
19. Frantzen, A.; Scheidtmann, J.; Frenzer, G.; Maier, W.F.; Jockel, J.; Brinz, T.; Sanders, D.; Simon, U. High-Throughput Method for the Impedance Spectroscopic Characterization of Resistive Gas sensors. *Angew. Chem. Int. Ed.* **2004**, *43*, 752–754. [CrossRef] [PubMed]
20. Siemons, M.; Koplin, T.J.; Simon, U. Advances in high throughput screening of gas sensing materials. *Appl. Surf. Sci.* **2007**, *254*, 669–676. [CrossRef]
21. Sanders, D.; Simon, U. High-Throughput Gas Sensing Screening of Surface-Doped $In_2O_3$. *J. Comb. Chem.* **2007**, *9*, 53–61. [CrossRef] [PubMed]
22. Li, C.; Zhang, S.; Hu, M.; Xie, C. Nanostructural ZnO based coplanar gas sensor arrays from the injection of metal chloride solutions: Device processing, gas-sensing properties and selectivity in liquors applications. *Sens. Actuators B Chem.* **2011**, *153*, 415–420. [CrossRef]
23. Wang, J.; Gao, S.; Zhang, C.; Zhang, Q.; Li, Z.; Zhang, S. A high throughput platform screening of ppb-level sensitive materials for hazardous gases. *Sens. Actuators B Chem.* **2018**, *276*, 189–203. [CrossRef]
24. Deng, Q.; Gao, S.; Lei, T.; Ling, Y.; Zhang, S.; Xie, C. Temperature & light modulation to enhance the selectivity of Pt-modified zinc oxide gas sensor. *Sens. Actuators B Chem.* **2017**, *247*, 903–915.
25. Li, D.; Lei, T.; Zhang, S.; Shao, X.; Xie, C. A novel headspace integrated E-nose and its application in discrimination of Chinese medical herbs. *Sens. Actuators B Chem.* **2015**, *221*, 556–563. [CrossRef]
26. Du, Y.; Gao, S.; Mao, Z.; Zhang, C.; Zhao, Q.; Zhang, S. Aerobic and anaerobic $H_2$ sensing sensors fabricated by diffusion membranes depositing on Pt-ZnO film. *Sens. Actuators B Chem.* **2017**, *252*, 239–250. [CrossRef]
27. Pardo, A.; Cámara, L.; Cabré, J.; Perera, A.; Cano, X.; Marco, S.; Bosch, J. Gas measurement systems based on IEEE1451.2 standard. *Sens. Actuators B Chem.* **2006**, *116*, 11–16. [CrossRef]
28. Gutierrezosuna, R. Pattern analysis for machine olfaction: A review. *IEEE Sens. J.* **2002**, *2*, 189–202. [CrossRef]

29. Geoffrey, M.; Kim-Anh, D.; Christophe, A. *Analyzing Microarray Gene Expression Data*; Wiley: New York, NY, USA, 2004. [CrossRef]
30. Macías, M.M.; Agudo, J.E.; Manso, A.G.; Orellana, C.J.G.; Velasco, H.M.G.; Caballero, R.G. A compact and low cost electronic nose for aroma detection. *Sensors* **2013**, *13*, 5528–5541. [CrossRef] [PubMed]

*micromachines*

MDPI

*Article*

# Fabrication and Packaging of CMUT Using Low Temperature Co-Fired Ceramic

**Fikret Yildiz [1,2,*], Tadao Matsunaga [3] and Yoichi Haga [3]**

[1] Graduate School of Engineering, Tohoku University, 6-6 Aza-Aoba, Aramaki Aoba-ku, Sendai 980-8579, Japan
[2] Faculty of Engineering, Hakkari University, Hakkari 30000, Turkey
[3] Graduate School of Biomedical Engineering, Tohoku University, 6-6 Aza-Aoba, Aramaki Aoba-ku, Sendai 980-8579, Japan; matsunaga@tohoku.ac.jp (T.M.); haga@tohoku.ac.jp (Y.H.)
* Correspondence: yildizfkrt@gmail.com; Tel.: +90-438-212-1212

Received: 11 October 2018; Accepted: 24 October 2018; Published: 27 October 2018

**Abstract:** This paper presents fabrication and packaging of a capacitive micromachined ultrasonic transducer (CMUT) using anodically bondable low temperature co-fired ceramic (LTCC). Anodic bonding of LTCC with Au vias-silicon on insulator (SOI) has been used to fabricate CMUTs with different membrane radii, 24 μm, 25 μm, 36 μm, 40 μm and 60 μm. Bottom electrodes were directly patterned on remained vias after wet etching of LTCC vias. CMUT cavities and Au bumps were micromachined on the Si part of the SOI wafer. This high conductive Si was also used as top electrode. Electrical connections between the top and bottom of the CMUT were achieved by Au-Au bonding of wet etched LTCC vias and bumps during anodic bonding. Three key parameters, infrared images, complex admittance plots, and static membrane displacement, were used to evaluate bonding success. CMUTs with a membrane thickness of 2.6 μm were fabricated for experimental analyses. A novel CMUT-IC packaging process has been described following the fabrication process. This process enables indirect packaging of the CMUT and integrated circuit (IC) using a lateral side via of LTCC. Lateral side vias were obtained by micromachining of fabricated CMUTs and used to drive CMUTs elements. Connection electrodes are patterned on LTCC side via and a catheter was assembled at the backside of the CMUT. The IC was mounted on the bonding pad on the catheter by a flip-chip bonding process. Bonding performance was evaluated by measurement of bond resistance between pads on the IC and catheter. This study demonstrates that the LTCC and LTCC side vias scheme can be a potential approach for high density CMUT array fabrication and indirect integration of CMUT-IC for miniature size packaging, which eliminates problems related with direct integration.

**Keywords:** capacitive micromachined ultrasonic transducers (CMUT); low temperature co-fired ceramic (LTCC); LTCC side via; indirect packaging

## 1. Introduction

The capacitive micromachined ultrasonic transducer (CMUT) is an advanced ultrasonic transducers technology and is based on a micro electro mechanical systems (MEMS). The simple structure of CMUT consists of a micromachined membrane suspended over a cavity, a fixed bottom electrode, and a top electrode [1,2]. It has attracted scientists and researchers in this field in recent years. There are several studies related to numerical and analytical methods of CMUT in addition to fabrication [3–9]. First generation CMUTs were fabricated using the surface micromachining/sacrificial layer releasing method [10]. This method includes several depositions and etching steps. Cavities under the membrane are obtained by selective etching of the sacrificial layer through etching holes that are patterned on the membrane and this reduces, however, the active area of membrane (fill factor). The membrane is deposited over the sacrificial layer and an additional deposition step is required to seal cavities. Each of the deposition steps induces stress on the membrane [11–16]. Thus, precise control over membrane thickness is very critical because it determines the mechanical properties of the membrane (low internal stress, mechanical loss etc.). Moreover, the other common problem of sacrificial layer releasing for cavity formation is stiction which occurs after selective etching of the sacrificial layer. Capillary forces on the membrane during drying of water in the cavity push the membrane to the bottom substrate and break if the membrane is not sufficiently thick [17].

Wafer bonding was introduced as an alternative to surface micromachining and provides simplicity, flexibility, and superior control over fabrication processes and material selections [17]. In wafer bonding, a single silicon crystal is used as a membrane and vacuum sealed cavities are achieved without opening etch holes on the membrane, both of which directly translate into a high performance device with high fill factor [18]. Fusion bonding and anodic bonding are mostly preferred wafer bonding methods for 1D/2D CMUT fabrication among other bonding techniques due to the advantages of bond strength, reliability, and hermiticity [19–24]. However, high bonding temperature, and the flat and clean bonding surface requirement are limitations of fusion bonding [25]. Anodic bonding, on the other hand, is a promising candidate for CMUTs fabrication and packaging (electronic integration) due to low temperature process compatibility. A CMUT uses all the benefits of advanced MEMS technology; however, it still needs improvements to show comparable performance to its piezoelectric counterpart in terms of sensitivity and output pressure. Due to small capacitance, CMUTs are sensitive to parasitic capacitance and have a low SNR (signal to noise ratio) value [26–28]. Low output pressure is other concern with CMUT performance. For ultrasound imaging and therapeutic applications, high SNR and output pressure are the main requirements as well as high dynamic range and low cross coupling between transducer elements [29–32]. To do this, the active area of the vibrating membrane would be increased and parasitic effects should be minimized. Direct integration of CMUT and front-end electronics (3D integration) is highly desirable to increase SNR and output pressure, but also reduction of parasitic effects. Thus, through-wafer interconnects are needed and electrical contact pads have to be located at the backside of the CMUT for 3D packaging and to provide communication between CMUT elements and the IC chip. Several materials and methods have been under investigation to show CMUT packaging with electronics. Earlier through-wafer interconnects efforts were widely through silicon via (TSV) [33]. The TSV process begins with vias opening on silicon substrate by deep reactive ion etching (DRIE) and then thermal oxidation of substrate for insulation. The next step is filling vias with a conductive material such as polysilicon which serves as conductor between the front side (CMUT) and backside (bonding pad) of the wafer. These TSV processes induce stress on the silicon substrate and require an additional polishing step to achieve a bondable surface for fusion bonding [34]. Alternatively, the through trench isolation approach has been announced to eliminate drawbacks related with the TSV method [35]. For example, a process has recently been reported for the fabrication of a CMUT array with isolation trenches using anodic bonding [36]. This study proposed a simple interconnects formation without through-wafer via. To date, the majority of works have focused on through-silicon vias (TSV), however, parasitic capacitance is an issue for such architecture. Using dielectric material in the form of through-glass vias (TGV) rather than Si can eliminate these undesired effects and low surface roughness is not needed for bonding [37,38]. Via formation and metallization of glass are not an easy and simple task although promising results of CMUT fabrication using Through Glass via (TGV) have been shown [39]. An alternative material called anodically bondable low temperature co-fired ceramic (LTCC) has been developed, which has been widely used for die level or wafer level MEMS packaging over past years. A narrower via pitch fabrication is easier than when using a glass substrate, and also LTCC allows freedom in via design [40–43]. Recently, SOI-LTCC anodic bonding has been announced for CMUT fabrication [24,44,45]. In these studies, CMUTs were built directly on open tool and customized LTCC substrate. Fabricated devices were electromechanically characterized for resonance frequency in air and immersion medium. Initial results showed that LTCC is one of the potential candidates for CMUT fabrication and hybrid integration with electronics. Moreover, LTCC has via and vertical interconnects which enables lateral side via architecture (indirect packaging) for electronic integration with IC. This is highly desirable for small size CMUT packaging, for example, tube shaped packaging of CMUT to visualize the narrower part of the vessel (intravascular imaging). In other words, lateral side and backside integration of CMUT with electronics are possible with LTCC substrate [46]. All aforementioned advantages of LTCC might provide high-density CMUT array fabrication and 3D packaging for different applications.

In this study, a custom designed LTCC wafer was used for CMUT fabrication and packaging. Bottom electrodes were directly built on LTCC via and high conductive silicon was used as top electrode and cavity formation. Anodic bonding of LTCC-SOI substrate was the final step of the fabrication process. A novel packaging process was introduced by using lateral side vias of LTCC that were achieved by micromachining of fabricated CMUT device. This packaging process refers to indirect integration of CMUT and IC using an intermediate material (catheter). Hexagonal shaped CMUTs with lateral side via were assembled with a catheter. ICs were mounted on the catheter following patterning of connection and contact pads using flip-chip bonding. Flip-chip bonding performances were evaluated and compared with similar studies in literature. Finally, the pros and cons of fabrication and packaging of LTCC based CMUTs were evaluated and discussed.

## 2. Materials and Methods

### 2.1. Fabrication

LTCC is a substrate made of a mixture of ceramic powder known as green sheet and a glass powder. Via fabrication is based on the following steps: (1) punching of green sheet for via hole formation, (2) screen printing of vias and interconnects (lateral wiring), and (3) stacking and firing of green sheet, respectively [42,43], as illustrated in Figure 1-(I). LTCC used in this study consists of vias with a diameter of 60 μm. However, LTCC has a 30 μm fabrication error. Lateral wiring (interconnects) and vias are made of conductive materials (Au) and provide electrical connections between the top and bottom of the LTCC substrate. LTCC and SOI substrates with a size of 2 cm × 2 cm were used for CMUT fabrication. The CMUT fabrication process was briefly summarized in Figure 1-(II). Table 1 shows CMUT fabrication parameters. In this micromachining process, the SOI wafer was firstly cleaned with Piranha solution ($H_2SO_4$:$H_2O_2$ = 2:1) to remove organics and contaminants. A thin layer of Au/Cr (20/30 nm) was sputtered on both sides of the LTCC wafer as a mask layer for wet etchant, and positive photoresist (PMER P-LA900, Tokyo Ohka Kogyo Co., Ltd., Kanagawa, Japan) with a thickness of 8 μm was spin coated for the lithography process. LTCC substrate was then exposed to wet etchant (HF:$H_2SO_4$ = 85:15) to obtained porous LTCC via. For 30 s etching time, etching depth and diameter of LTCC via were 10 μm and 150 μm, respectively. Bottom electrodes made of Au/Pt/Cr (70/30/20 nm) were patterned on the remaining LTCC via by a lift-off process after removal of resist and metals. Positive photoresist of 2 μm thick (OFPR-800 LB 200 cP, Tokyo Ohka Kogyo Co.,Ltd, Kanagawa, Japan) was used for photolithography and the lift-off process of the bottom electrode formation (Figure 2a). In the view of the SOI substrate, 0.4 μm circular shape cavities were micromachined on the high conductive Si part of SOI by reactive ion etching (RIE) with SF6 gas and Si used as a top electrode. Cr/Au metal bump with a thickness of 1.2 μm were sputtered on Si substrate for bonding with wet etched LTCC via during anodic bonding (Figure 2b). Anodic bonding of LTCC-SOI substrates was completed under a high vacuum condition at 380 °C. We repeated anodic bonding 27 times, resulting in 243 single CMUTs die. Because 2 cm × 2 cm LTCC includes nine different via designs comprising a linear and circular array. Au-Au bonding of wet etched LTCC vias and bumps forms an electrical connection between the top and bottom of the bonded sample [40,41]. The undesired part of bonded LTCC-SOI substrates (handling layer-300 μm in thickness) was then removed by deep reactive ion etching (DRIE) following by contact pads formation at the backside of LTCC (Figure 2c). Fabricated CMUTs were characterized in air and immersion medium. More details about LTCC-SOI bonding process flow have been given in our previous studies [44].

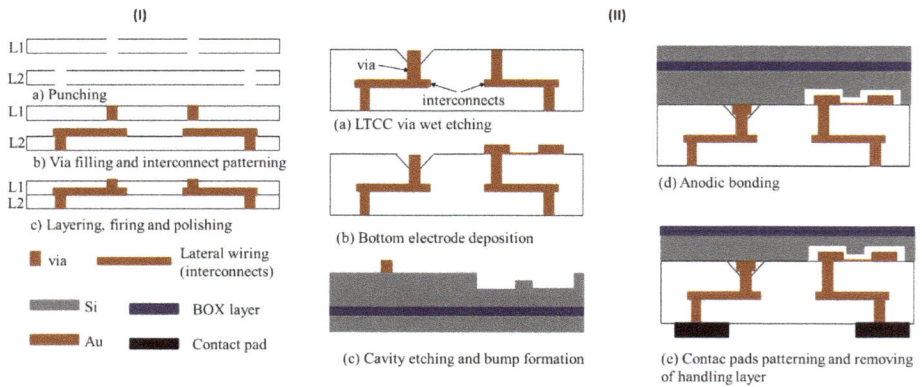

**Figure 1.** (**I**) Low temperature co-fired ceramic (LTCC) fabrication: (**a**) Via hole opening by punching; (**b**) Filling hole and interconnects patterning; (**c**) Layering and firing. (**II**) Capacitive micromachined ultrasonic transducer (CMUT) fabrication process: (**a**) Porous via formation by LTCC wet etching; (**b**)Bottom electrode deposition; (**c**) Cavity etching and Au-bump deposition; (**d**) Anodic bonding and (**e**) Contact pad patterning and handling layer removing.

**Table 1.** The physical parameters of fabricated CMUT devices.

| Parameters | Value |
|---|---|
| Membrane diameter (μm) | 48, 50, 72, 80, 120 |
| Membrane thickness (μm) | 2.6 |
| Cavity depth (μm) | 0.4 |
| Electrode thickness (nm) | 120 |
| Number of elements | 25, 34, 54 |

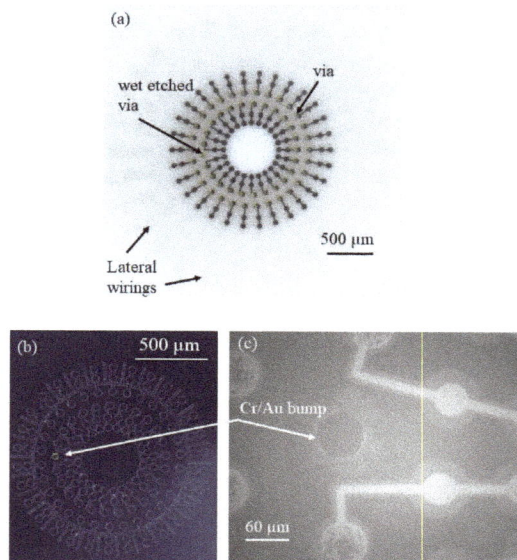

**Figure 2.** Preparation of LTCC and Si substrate for anodic bonding (Top view). (**a**) Bottom electrodes on dual ring LTCC with a number of 25 inners and 30 outer via; (**b**) Bumps and cavities on Si and (**c**) IR view of device patterns of CMUT after handling layer removing.

## 2.2. Packaging

Packaging process flow of indirect CMUT-IC packaging is described in this section of paper. This novel packaging process uses the lateral side via of LTCC rather than the backside of substrate for packaging. Initial results and more details have been found in our previous research in [45,46]. According to [46], indirect connection of CMUT with lateral side via and IC circuits was proposed through patterned electrodes on the LTCC side via and catheter. The catheter is made of a biocompatible solid polyimide substrate with a size of 3 mm × 3 mm × 20 mm. This process consists of four different steps: (1) machining of fabricated CMUT and catheter, (2) assembly of CMUT and catheter, (3) electrode and contact pad patterning on both substrate, and (4) IC mounting on catheter by flip-chip bonding. Lateral side vias were obtained by cutting the CMUT device in hexagonal (Φ: 2.4 mm) and rectangular shapes (Φ: 3 mm) using the dicing machine as illustrated in Figure 3a. Diamond blades 1 mm and 0.1 mm thick were used for micromachining of the fabricated CMUT and catheter, respectively. The micromachined catheter has three different regions: first planar surface (length: 3 mm), taper (length: 2 mm), and second planar surface (length: 10 mm). The taper was formed with a 0.05 mm dicing pitch although the other part of the catheter was diced with a 0.1 mm dicing pitch. The taper depth was the sum of IC chip thickness and contact pads thickness of both flip-chip bonded samples (IC and catheter). The assembly process follows micromachining and is the mounting of the catheter to the backside of the CMUT using epoxy adhesive. Alignment of the catheter and CMUT was achieved with a set of sample holders and a fixer. They were fabricated with a 3D printer (Agilista-3000, Keyence Co., Osaka, Japan).

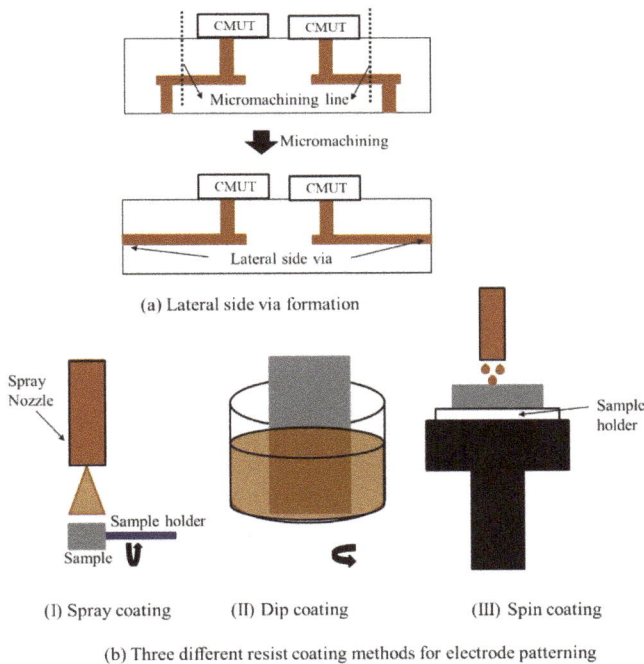

(a) Lateral side via formation

(I) Spray coating    (II) Dip coating    (III) Spin coating

(b) Three different resist coating methods for electrode patterning

**Figure 3.** CMUT packaging process flow. (**a**) Side via formation and (**b**) electrode patterning methods.

The first sample holder was designed and fabricated for CMUT and the other was the catheter. Holder and fixer have alignment holes and pins for assembly process. Samples were put inside the holders and high temperature resistance adhesive (EPO-TEK® 353ND, Epoxy Technology Inc., Billerica, MA, USA) was then used for assembly. Alignment of samples was achieved with alignment holes and pins and samples were strictly put together by pushing the fixer at 100 °C curing temperature for around 1 h. Three different coating methods were used for connection electrodes, wiring pads, and bonding pads patterning on side via and catheter after assembly: spray coating, dip coating, and spin coating as shown in Figure 3b. The Nonplanar exposure system which consists of a UV spot laser and a computer-controlled multiaxial stage used resist patterning for spray coating and dip coating [47,48]. However, the spin coating method was preferred to electrode patterning using a planar exposure system. Lithography process results of three coating methods showed that spin coating and the planar exposure system were the best fitted methods for electrode and contact pad patterning on assembled samples. Therefore, electrodes and contact pads were patterned by spin coating and planer exposure system using a contact mask aligner (Ma8, Suss MicroTec KK, Kanagawa, Japan). Electrodes and contact pads deposition were achieved using the following steps: 1.2 µm thick Cr/Au electrodes and alignment marks were firstly patterned on assembly by a lift-off process using positive photoresist (OFPR- 800 LB 200 cP, Tokyo Ohka Kogyo Co., Ltd., Kanagawa, Japan). These electrodes provide a connection between LTCC side via and IC bonding pads that are on the catheter. Wiring pads and bonding pads were then electroplated after patterning of positive photoresist (PMER P-LA900, Ohka Kogyo Co., Ltd., Kanagawa, Japan). Wiring pads were designed and deposited for measurement of the resistance between the flip-chip bonding pads. The widths of the bonding pads and connection electrodes have widths of 70 µm and 20 µm and pitches of 140 µm and 150 µm, respectively. It was measured that thickness of deposited electrode and pads were 5 µm. A dummy IC chip was used for flip-chip bonding. Silicon-on-insulator (SOI; 3 µm/50 nm/300 µm) substrate was preferred as a dummy IC and includes eight bonding pads. Au/Cr bonding pads (80/200 nm) were first patterned and then 50 µm thick Au bumps were formed on the Au/Cr pads using a wire bonder (7700 West Bonder, West Bond Inc., Anaheim, CA, USA). Finally, an IC chip was mounted on the second planar surface of the catheter using a flip-chip bonder (FINEPLACER®lambda, Finetech GmbH & Co. KG, Berlin, Germany). A bonding force of 25 N was applied for 3 min and heated to 380 °C. Flip-chip bonding results were evaluated by resistance measurement of bonding pads [46].

## 3. Results

Anodic bonding quality evaluation is needed to show functionality of fabricated devices. The first bond quality evaluation of the bonded sample was dicing of samples into small pieces (0.6 cm × 0.6 cm) using a dicing machine (DAD 522, DISCO Co., Tokyo, Japan). Bonding is considered as successful when bonded samples stayed together. In order to inspect the bonding quality more accurately, the fabricated devices were tested with three additional different measurements: (1) Visual inspection of bonded samples to detect misalignment using IR camera, (2) impedance analyzer for the measurement of admittance (G (conductance)-B (susceptance)) as a function of the frequency, and (3) static membrane deflection by topography measurement system (TMS) (Polytec Japan, Kanagawa, Japan) to show hermiticity of the sealed cavity. Visual inspections of bonded samples were firstly tested using the IR camera. Top views of four different CMUT devices obtained by IR camera are shown in Figure 4a,c,d. Our results showed that there was a misalignment between the top and bottom electrode of CMUTs with membrane diameter of 48 µm, 50 µm, and 80 µm. In addition to the LTCC via error (30 µm), an approximately 20 µm mechanical error was measured during samples preparation (alignment, clamping etc.) before the bonding process. It was assumed that the mechanical error related with the bonding machine was the reason for misalignment in addition to the via fabrication error. Gold diffusion into silicon was also observed due to short contact of the Si membrane and bottom electrode made of Au/Pt/Cr as a result of misalignment (Figure 4b). It was noted that the CMUT cell with a dimension of 120 µm has no misalignment as shown in Figure 4d. Misalignment and a short connection between the top and bottom electrode is also confirmed by complex admittance measurement. Conductance, G ($\omega$), refers to the real part and susceptance, B ($\omega$), presents the imaginary part of complex admittance. Lumped equivalent circuit and values of circuit parameters can be obtained by plotting B ($\omega$) versus G ($\omega$) as described in [49].

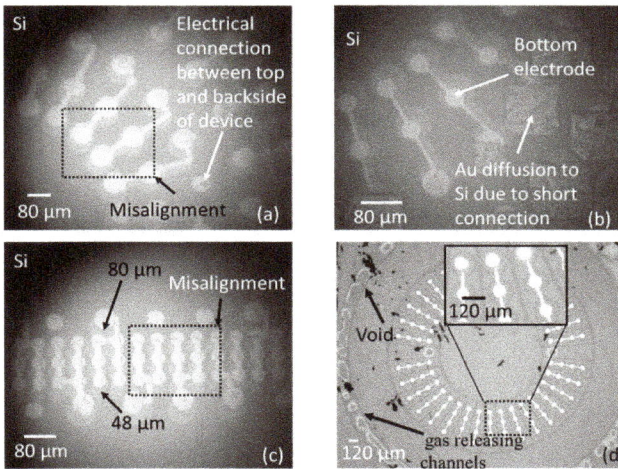

**Figure 4.** Three CMUT devices with different membrane size. (**a**) 80 μm, (**b**) Au diffusion in Si membrane, (**c**) linear array CMUT with a 48 μm membrane size, d and (**d**) CMUT ring array with a 120 μm membrane size.

Complex admittance measurement by impedance analyzer (HP4194A, Hewlett Packard, Co., Palo Alto, CA, USA) was employed to obtain G (conductance)-B (susceptance) plot of fabricated devices. It simply gives an idea about the characteristics (equivalent circuit) of fabricated devices that can be a resistor, capacitor, inductor, or a combination of three electronic circuit elements. Resistance and frequency value of fabricated devices were derived by plotting the imaginary part of the admittance, B (ω), versus the real part, G (ω). G–B plot of four different CMUT designs and their equivalent circuits are shown in Figure 5 and the inset of Figure 5, respectively. Fabrication results showed that equivalent circuits of three fabricated devices (40 μm, 50 μm, and 80 μm) consist of a capacitor with a series resistor (R1) and a parallel resistor (R2). For CMUT with a 72 μm membrane size, a capacitor is the only parameter of equivalent circuits as expected. Logarithmical curve fitting was applied to find the best suited function for the first three designs, and linear curve fitting matches the data of the last design (72 μm). Lastly, TMS was used to measure static deflection of the CMUT membrane under atmospheric pressure as shown in Figure 6. It was observed that the membrane deflection profile of the Si membrane was in an upward direction. In contrast to the CMUT membrane, deflection of the Si part around the wet etched LTCC via was downward, as shown in Figure 6c,d. Mechanical and electrical characterization of CMUTs in air were determined by resonance frequency and impedance measurements. Resonance frequency of a device in air was measured using a vibrometer (UHF-120, Polytec Japan, Kanagawa, Japan), and a network analyzer (MS4630B, Anritsu, Co., Morgan Hill, CA, USA) was used for impedance measurement. CMUT with a 120 μm membrane size was used for experimentation. The measured maximum membrane displacement in air was 10.3 pm at 2.88 MHz under excitation with a 7 Vpp AC signal without DC bias voltage. A finite element model of a CMUT cell was constructed in COMSOL Multiphysics (COMSOL®version 5.2, COMSOL, Inc., Burlington, MA, USA) software, coupling the structural mechanics subdomain and the electrostatics subdomain to compare experimental and numerical results.

The 2D electromechanical coupling model was used. The free triangular mesh and defaults parameters were set for calculation. The Minimum feature size was 0.054 μm. The fixed mesh was applied to LTCC and the bottom electrode when modeling. However, Si membrane and cavities were free to deform. Squeezed film damping in sealed cavity was omitted for modelling, because it was analytically proved from a previous study that the presence of air does not cause any squeeze film damping for flexural membrane [50]. Resonance frequency and maximum membrane displacement were obtained as 1.5755 MHz and 2.42 pm according to the numerical analysis as shown in Figure 7 [44].

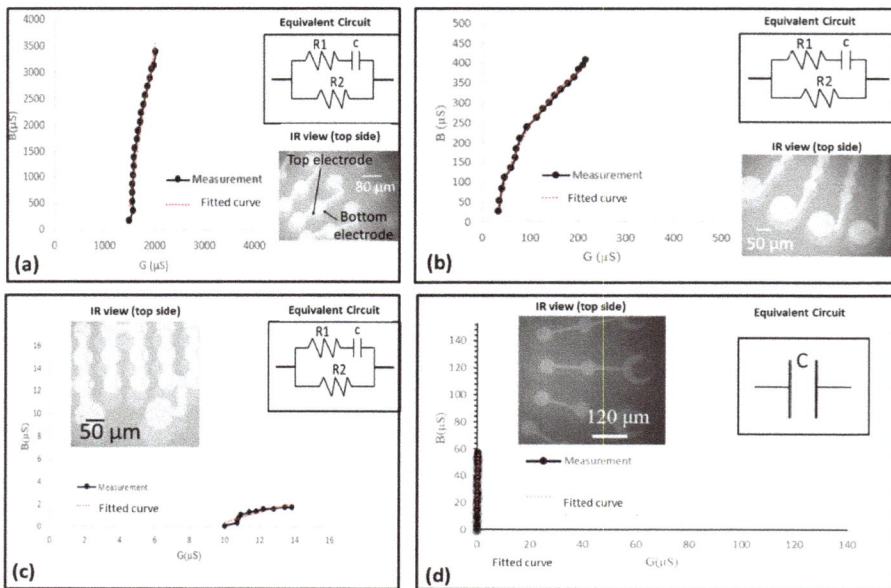

**Figure 5.** Complex admittance plot of four different CMUT devices: (**a**) 80 μm; (**b**) 50 μm; (**c**) 48 μm; and (**d**) 72 μm.

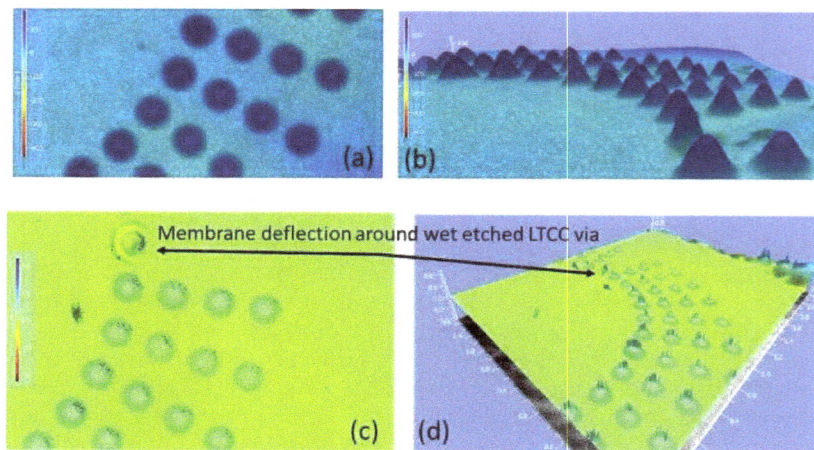

**Figure 6.** Membrane deflection of successfully bonded sample. Membrane deflection of CMUT with a 120 μm membrane size. (**a**) Top view (deflection: 162.5 nm); (**b**) 3D view (deflection: 174.5 nm). Hermitically sealed cavity around wet etched LTCC via; (**c**) Top view and (**d**) 3D view.

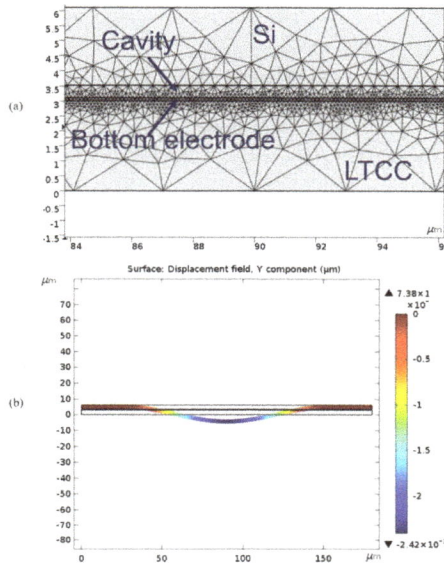

**Figure 7.** (**a**) Mesh information of 2D FEM model of a CMUT cell and (**b**) Vertical displacement under 7V DC bias.

Experimental results of the impedance measurement are shown in Figure 8a. Applied DC voltages are changed from 10 V to 40 V and resonance span is changed from 1–4 MHz. The second experimental setup is the pitch-catch setup where one of the transducers transmits an acoustical signal, and this signal was measured by a hydrophone (TC4038, Teledyne RESON Inc., Thousand Oaks, CA, USA) placed at a distance from the CMUT surface in water as shown in Figure 8b. This hydrophone has a frequency ranging up to 20 MHz. However, we could not observe any peak around the resonance frequency. Moreover, CMUT devices used in electrical and acoustical measurement were not successfully driven by different AC and DC voltages while device fabrication was successful.

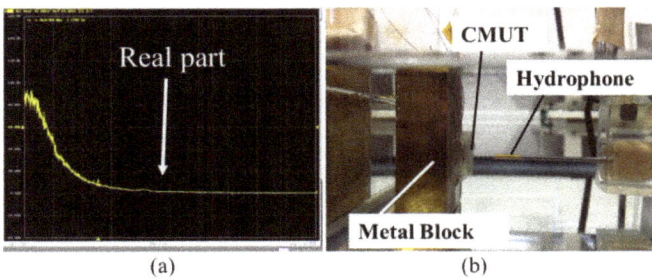

**Figure 8.** Device characterizations. (**a**) Electrical measurement by impedance analyzer and (**b**) acoustic measurement in water by hydrophone.

Surface roughness measurement of catheter after micromachining were the first experimental results of packaging. It was measured that the surface roughness of the lateral side of the first planar surface was about to 20–30 μm. When these surfaces were polished to reduce surface roughness, the shape of catheter turned into a circular shape. When the assembly process is considered, epoxy adhesive that is compatible with low temperatures is not a good choice for the assembly of CMUT and catheter (Figure 9). Experimental results showed that amount of adhesive should be confirmed before the assembly process because squeezed adhesive from the interface of the bonded area covered the side vias and electrodes on catheters and this prevents next electrodes patterning. Experimental results of planar and nonuniform exposure systems were evaluated in terms of resist patterning and electrode deposition after the assembly of CMUT and catheter. A requirement for different focusing points of the laser (due to planar and taper of catheter) was unable to achieve successful photoresist patterning on the catheter using the nonplanar system. LTCC side vias were used as reference points for alignment and laser exposure for nonplanar system. However, electrodes patterning on the LTCC side via and nonuniform catheter surface is a very complex and difficult process using nonuniform exposure systems. The planar exposure system, therefore, was preferred for electrodes deposition on the catheter, even on the taper. A longer exposure time is required to resist patterning on the taper than for planar surfaces of the catheter before electrode deposition. Electrodes and contact pads were electroplated with a thickness of 5 μm. Contact resistance of bonding pads after flip-chip bonding was measured using 4-wire measurement setup. Resistance of each bump measured around 2 μm although the theoretical value of a single bump was about 0.25 μm [46]. These results are considerably lower than in previous studies in literature [33]. In addition to evaluation of flip-chip bonding success, resistance between electrodes on the side via and catheter was measured with a 100% yield [46]. Figure 10 shows summary of successful packaging process flow from micromachining of the CMUT and catheter to IC mounting.

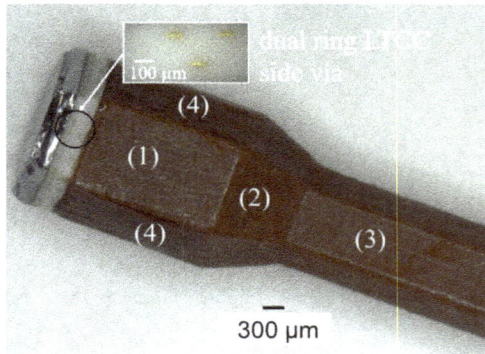

**Figure 9.** Assembled hexagonal shaped CMUT and catheter. (1) is the first planer surface of catheter; (2) taper; (3) 2nd planer surface; and (4) lateral side of catheter.

(a) Micromachined cMUTs

(b) Assembly and electrode patterning

(c) IC mounting using flip-chip bonding

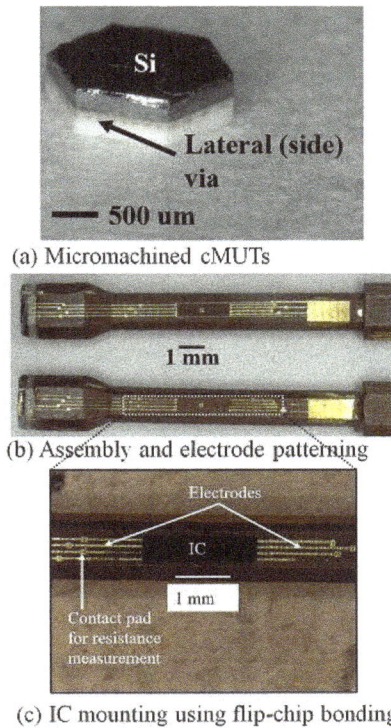

**Figure 10.** Indirect CMUT-IC packaging process using LTCC side via. (**a**) Hexagonally micromachined CMUT; (**b**) Electrodes and contact pads on catheter after mounting of catheter at the backside of CMUT and (**c**) IC mounting on catheter.

## 4. Discussion

In this section of paper, drawbacks, limitations, possible reasons behind undesired results of CMUT fabrication and packaging were discussed. The IR picture also revealed that no void and bubbles on the active area of the bonded surface were observed with the help of gas releasing channels. These channels were patterned between array elements and the circumference of device as shown in Figure 4d. Thus, we can say that the bonding strength was enough and voids were only visible in the gas releasing channels without significant effect on bonding strength. From the complex admittance results of three unsuccessful CMUTs, we assumed that short connections and particles on the bottom electrode that remained after the fabrication process are responsible for a parallel and a series resistor to CMUT device (capacitor), respectively. TMS results prove that the membrane deflection is upward, however, membrane displacement over the Au-Au bonded area, which was used for the electrical connection between the top and bottom surface, has a downward direction with 90 nm displacement (Figure 6c,d). Three possible reasons responsible for membrane deflection were investigated. These are gas trapped inside the cavity, the residual thermal stress on the Si surface during bonding, and TEC (thermal expansion coefficient) differences between Si and the LTCC substrate. Our previous study showed that thermal stress on the Si membrane during bonding was assumed to be a major factor behind the membrane deflection in an upward direction based on numerical analysis of thermal stress on silicon [44]. According to [44], membrane displacement due to gas trapped inside the cavity and TEC mismatch can be ignored. Considering Si membrane displacement due to high thermal stress, it can be concluded that cavities of fabricated devices were successfully sealed, however, without a vacuum due to outgassing during bonding. Previously, acoustical characteristics of the open tool LTCC-based CMUT device were shown and membrane displacement in the air was measured to be 10 times higher than in water [24]. According to our experimental results, membrane displacement in air was around 10.3 pm and the membrane displacement of our device in water should be 0.1 pm, when considering the experimental results of [24]. This displacement is very small and, thus, the output pressure of the device might not be within the range of hydrophone sensitivity for immersion measurement (acoustic). After handling layer removal, it was also observed that the silicon membrane had been removed and the membrane had collapsed to the bottom electrode in some cases, as shown in Figure 11a,b. This structure was repeatedly observed from several bonded samples. These results proved that holes and cracks at the surface of the vibration membrane made it unable to operate in the immersion medium. The Si membrane at this moment (after handling layer removing) might not be stiff enough to maintain its shape after handling layer removing. Nonlinear behavior of a Si membrane known as spring hardening can be another reason for unsuccessful device characterization due to high residual stress. Moreover, a low quality Si membrane due to overdamping in air and water can also prevent device operation. As a result, unsuccessful CMUT operation in air and water is possible due to aforementioned reasons related with the Si membrane. Because it was announced that theoretical modelling of the CMUT membrane with residual stress and cracks affects device performance (eigen frequency) [51–53], a thicker Si membrane can be a potential solution to eliminate cracks and holes on the membrane surface in addition to obtaining a stiffer membrane. To confirm this, Si membrane having 14 µm thickness and 80 µm diameter was used for anodic bonding with open tool LTCC. As shown in Figure 12, IR pictures of CMUT from the top side show no misalignment and cracks/holes on the membrane after anodic bonding.

The packaging process used in this was designed for indirect integration of CMUT with IC. Besides the advantages of CMUT packaging with LTCC, drawbacks and limitations of these new suggested methods should be considered for further improvements. From the packaging results, we can say that small size CMUT packaging is possible by using indirect connection of device and electronics rather than direct bonding. Excitation of a single CMUT array through two side vias (one is ground and other is hot electrode) might be easier than excitation of a single CMUT cell and dual ring array through multiple side vias due to a small via pitch. Moreover, it was assumed that insufficient heat flow, force, and bending of the catheter during flip-chip bonding were the main reasons behind the high contact resistivity of bump after bonding. Another micromachining method would be investigated or simple CMUT and catheter geometry would be used for electrode patterning and flip-chip bonding to gain more reliable results due to the difficulties of electrodes patterning on the nonuniform shape of the catheter. A square shaped CMUT and catheter, for example, can be a possible approach to drive a CMUT cell successfully from IC circuits patterned on a catheter. Because the surface roughness of the square shaped catheter was about the 2 µm, significantly lower than lateral side of hexagonal shape catheter (20–25 µm), the taper is no longer needed. Moreover, during the direct integration process, it is inevitable to prevent mechanical damages on a vibrating membrane of CMUT. To verify and validate mechanical damages on the active area of CMUT, CMUT was mounted on a dummy substrate (Pyrex glass) with bonding pads using flip-chip bonding by applying 25 N during 3 min. Cracks on the surface and deformed cells were observed after the bonding process (Figure 13a). LTCC side via approach for different CMUT shapes rather than a hexagon, therefore, can enable more functional and high performance CMUT fabrication and packaging. When a small size CMUT is required, we propose a packaging method where the connection between a square shaped CMUT and an IC circuit can be achieved using wire bonding and the flip-chip bonding by eliminating the taper on the catheter. This proposed packaging design is illustrated in Figure 13b.

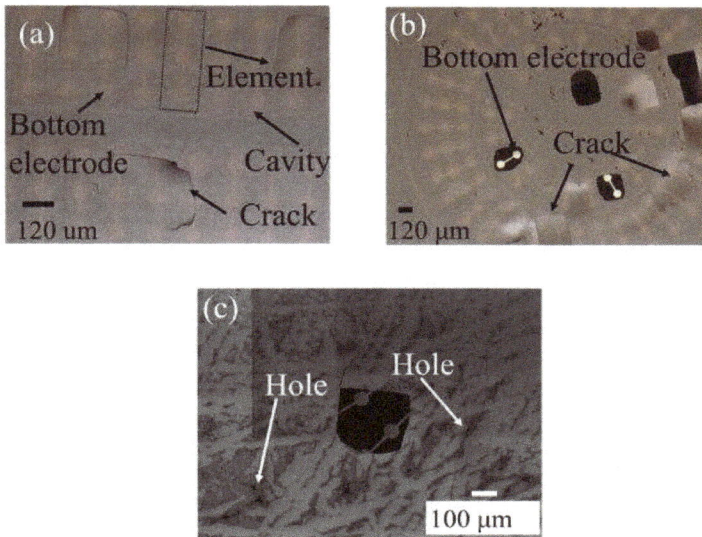

**Figure 11.** Cracks and holes observed at the CMUT membrane after handling-layer removal: (**a**) linear array; (**b**) dual ring array; and (**c**) SEM image of holes on the ring array CMUT surface after handling layer removing.

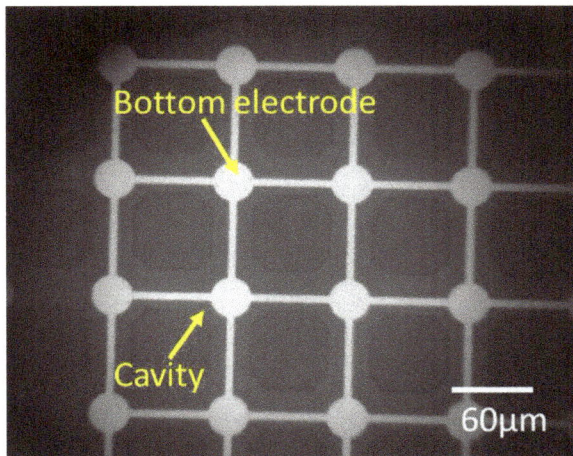

**Figure 12.** Anodic bonding result of 18 μm thick Si substrate and LTCC without cracks and holes on membrane.

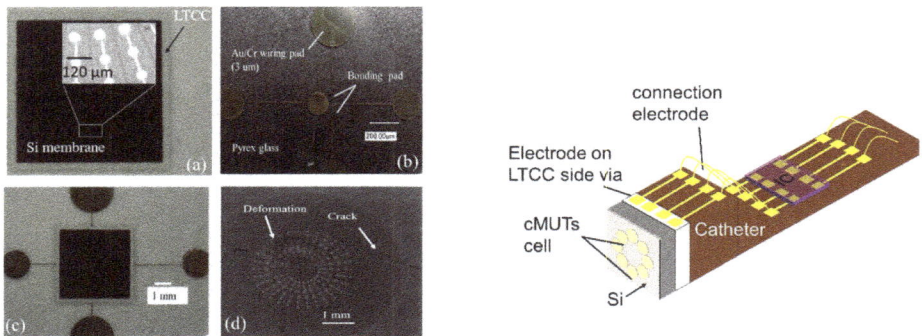

**Figure 13.** Direct integration of CMUT on dummy IC (**left**) and proposed indirect CMUT- IC connection scheme using flip-chip bonding and wire bonding (**right**).

## 5. Conclusions

In this paper, we have discussed various aspects of CMUT fabrication and packaging using LTCC substrate. Circular shaped CMUT cells with different sizes were successfully fabricated by LTCC-SOI anodic bonding. Infrared images and complex admittance plots were used to evaluate the bonding quality of dual, single ring array, and linear array CMUT. Device characteristics were investigated by obtaining equivalent circuits of devices derived from admittance plots. Our results showed that CMUT membrane size optimization does not easily achieve successful device fabrication due to via fabrication error and mechanical error of the bonding machine. It was found that a fabricated device is the only capacitor when a CMUT has a 72 µm membrane diameter. However, a resistive component was observed in the case of a CMUT with a diameter of 80 µm. Static membrane deflection at atmospheric pressure was measured to validate hermiticity of the cavity. The resonance frequency of the CMUTs with the 120 µm membrane diameters were measured at 2.88 MHz in air with a 10.3 pm displacement. Electrical and acoustical measurement of CMUTs in air and water were unsuccessful due to fabrication process related cracks and holes on the vibrating membrane. It is concluded that this caused a short connect between the top and bottom electrodes. These results indicated that the LTCC based CMUT might be suitable for air coupled applications such as gas sensing rather than immersion medium.

3D Integration of CMUT with an integrated circuit (IC) has been also investigated by using the lateral side via of LTCC. Micromachining, assembly, and electronic integration of the CMUT and catheter were presented. The LTCC side via was obtained by micromachining the CMUT into a hexagon. Connections between the CMUT and IC were achieved through electrodes patterned on a catheter that was mounted at the backside of CMUT. Electrode thickness was optimized to prevent disconnection between CMUT and IC. It was found that 5 µm electrode thickness was high enough to drive CMUT successfully. Contact resistance of flip-chip bonding was measured using a 4-wire measurement. 2 µm contact resistance was measured which is an acceptable range compared to previous studies. This indirect packaging technology might enable the integration of CMUT and integrated circuit (IC) for small sizes of ultrasonic systems.

**Author Contributions:** F.Y., T.M., and Y.H. designed and conducted the project. F.Y. performed the experiments. The data summary and writing of the article was mainly done by F.Y., T.M. and Y.H. gave suggestions and helpful discussion on the experiments and manuscript writing. All authors read and approved the final manuscript.

**Acknowledgments:** Part of this work was performed in the Micro/Nanomachining Research Education Center (MNC) of Tohoku University. This work was supported in part by Translational Research Network Program of The Ministry of Education, Culture, Sport, Science and Technology (MEXT) and the Creation of Innovation center for Advanced Interdisciplinary Research Areas Program of Japan Science and Technology Agency (JST). The authors gratefully acknowledge the NIKKO Company for design and fabrication of LTCC substrates.

**Conflicts of Interest:** The authors declare no conflict of interest.

# References

1. Khuri-Yakub, B.T.; Oralkan, O. Capacitive micromachined ultrasonic transducers for medical imaging and therapy. *J. Micromech. Microeng.* **2011**, *21*, 54004–54014. [CrossRef] [PubMed]

2. Oralkan, O.; Ergun, A.S.; Johnson, J.A.; Karaman, M.; Demirci, U.; Kaviani, K.; Lee, T.H.; Khuri-Yakub, B.T. Capacitive Micromachined Ultrasonic Transducers: Next-Generation Arrays for Acoustic Imaging. *IEEE Trans. Ultrason. Ferroelect. Freq. Control* **2002**, *49*, 1596–1610. [CrossRef]

3. Lohfink, A.; Eccardt, P.-C. Linear and nonlinear equivalent circuit modeling of CMUTs. *IEEE Trans. Ultrason. Ferroelectr. Freq. Control* **2005**, *52*, 2163–2172. [CrossRef] [PubMed]

4. Köymen, H.; Şenlik, M.N.; Alar, A.A.; Olcum, S. Parametric linear modeling of circular CMUT membranes in vacuum. *IEEE Trans. Ultrason. Ferroelectr. Freq. Control* **2007**, *54*, 1229–1239. [CrossRef] [PubMed]

5. Oleum, S.; Senlik, M.N.; Bayram, C.; Atalar, A. Design charts to maximize the gain-bandwidth product of capacitive micromachined ultrasonic transducers. *Proc. IEEE Ultrason. Symp.* **2005**, *4*, 1941–1944.

6. Bayram, C.; Olcum, S.; Senlik, M.N.; Atalar, A. Bandwidth improvement in a CMUT array with mixed sized elements. *Proc. IEEE Ultrason. Symp.* **2005**, *4*, 1956–1959.

7. Koymen, H.; Atalar, A.; Guler, S.; Koymen, I.; Tasdelen, A.S.; Unlugedik, A. Unbiased Charged Circular CMUT Microphone: Lumped-Element Modeling and Performance. *IEEE Trans. Ultrason. Ferroelectr. Freq. Control* **2018**, *65*, 60–71. [CrossRef] [PubMed]

8. Jallouli, A.; Kacem, N.; Bourbon, G.; le Moal, P.; Walter, V.; Lardies, J. Pull-in instability tuning in imperfect nonlinear circular microplates under electrostatic actuation. *Phys. Lett. Sect. A Gen. At. Solid State Phys.* **2016**, *380*, 3886–3890. [CrossRef]

9. Kacem, N.; Jallouli, A.; Walter, V.; Bourbon, G.; Lemoal, P.; Lardies, J. Nonlinear Dynamics of Circular Capacitive Micromachined Ultrasonic Transducers. In Proceedings of the 2015 IEEE SENSORS, Busan, Korea, 1–4 November 2015; pp. 1–4.

10. Haller, M.I.; Khuri-Yakub, B.T. A surface micromachined electrostatic ultrasonic air transducer. In Proceedings of the IEEE Ultrasonics Symposium, Cannes, France, 31 October–3 November 1994; pp. 1241–1244.

11. Jin, X.; Ladabaum, I.; Khuri-Yakub, B.T. The microfabrication of capacitive ultrasonic transducers. In Proceedings of the International Conference on Solid-State Sensors and Actuators, Chicago, IL, USA, 16–19 June 1997; pp. 437–440.

12. Zhang, Q.; Cicek, P.-V.; Allidina, K.; Nabki, F.; El-Gamal, M.N. Surface-Micromachined CMUT Using Low-Temperature Deposited Silicon Carbide Membranes for Above-IC Integration. *J. Microelectromech. Syst.* **2014**, *23*, 482–493. [CrossRef]

13. Ergun, A.S.; Cheng, C.-H.; Demirci, U.; Khuri-Yakub, B.T. Fabrication and Characterization of 1-Dimensional and 2-Dimensional Capacitive Micromachined Ultrasonic Transducer (CMUT) Arrays for 2- Dimensional and Volumetric Ultrasonic Imaging. In Proceedings of the Ocens '02 MTS/IEEE, Biloxi, MI, USA, 29–31 October 2002; Volume 4, pp. 2361–2367.

14. Ergun, A.S.; Zhuang, X.; Huang, Y.; Oralkan, O.; Yaralioglu, G.G.; Khuri-Yakub, B.T. Capacitive micromachined ultrasonic transducer technology for medical ultrasound imaging. In *Medical Imaging 2005: Ultrasonic Imaging and Signal Processing*; SPIE: Bellingham, WA, USA, 2005; Volume 5750, pp. 58–68.

15. Ergun, A.S.; Huang, Y.; Cheng, C.H.; Oralkan, O.; Johnson, J.; Jagannathan, H.; Demirci, U.; Yaralioglu, G.G.; Karaman, M.; Khuri-Yakub, B.T. Broadband Capacitive Micromachined Ultrasonic Transducer Ranging From 10 kHZ to 60 kHZ for Imaging Array and More. In Proceedings of the IEEE Ultrasonic Symposium, Munich, Germany, 8–11 October 2002; pp. 1039–1043.

16. Jin, X.; Değertekin, F.L.; Calmes, S.; Zhang, X.J.; Ladabaum, I.; Khuri-Yakup, B.T. Micromachined Capacitive Transducer Arrays for Medical Ultrasound Imaging. In Proceedings of the IEEE Ultrasonic Symposium, Sendai, Japan, 5–8 October 1998; pp. 1877–1880.

17. Ergun, A.S.; Huang, Y.; Zhuang, X.; Oralkan, O.; Yaralioglu, G.G.; Khuri-Yakup, B.T. Capacitive Micromachined Ultrasonic Transducers: Fabrication Technology. *IEEE Trans. Ultrason. Ferroelectr. Freq. Control* **2005**, *52*, 2242–2258. [PubMed]

18. Huang, Y.; Ergun, A.S.; Haeggstrom, E.; Badi, M.H.; Khuri-Yakub, B.T. Fabricating Capacitive Micromachined Ultrasonic Transducers With Wafer-Bonding Technology. *J. Microelectromech. Syst.* **2003**, *12*, 128–137. [CrossRef]

19.  Yamaner, F.Y.; Zhang, X.; Oralkan, Ö. Fabrication of anodically bonded capacitive micromachined ultrasonic transducers with vacuum-sealed cavities. In Proceedings of the IEEE International Ultrasonics Symposium, Chicago, IL, USA, 3–9 September 2014; pp. 604–607.

20.  Tsuji, Y.; Kupnik, M.; Khuri-Yakub, B.T. Low temperature process for CMUT fabrication with wafer bonding technique. In Proceedings of the 2010 IEEE International Ultrasonics Symposium, San Diego, CA, USA, 11–14 October 2010; pp. 551–554.

21.  Bayram, B. Diamond-based capacitive micromachined ultrasonic transducers. *Diam. Relat. Mater.* **2012**, *22*, 6–11. [CrossRef]

22.  Midtbø, K.; Rønnekleiv, A.; Wang, D.T. Fabrication and characterization of CMUTs realized by wafer bonding. In Proceedings of the 2006 IEEE Ultrasonics Symposium, Vancouver, BC, Canada, 2–6 October 2006; Volume 1, pp. 934–937.

23.  Bellaredj, M.; Bourbon, G.; Walter, V.; le Moal, P.; Berthillier, M. Anodic bonding using SOI wafer for fabrication of capacitive micromachined ultrasonic transducers. *J. Micromech. Microeng.* **2014**, *24*, 025009. [CrossRef]

24.  Hiroshima, M.; Matsunaga, T.; Mineta, T.; Haga, Y. Capacitive Micromachined Ultrasonic Transducers Using Anodically Bondable Ceramic Wafer with Through-Wafer Via. *IEEJ Trans. Sens. Micromach.* **2014**, *134*, 333–337. [CrossRef]

25.  Oberhammer, J. *Novel RF MEMS Switch and Packaging Concepts*; Royal Institute of Technology (KTH): Stockholm, Sweden, 2004.

26.  Mills, D.M. Medical Imaging with Capacitive Micromachined Ultasound Transducer (CMUT) Arrays. In Proceedings of the IEEE Ultrasonics Symposium, Montreal, QC, Canada, 23–27 August 2004; Volume 1, pp. 384–390.

27.  Percin, G.; Khuri-Yakub, B.T. Piezoelectrically Actuated Flextensional Micromachined Ultrasound Transducers—II: Fabrication and Experiments. *IEEE Trans. Ultrason. Ferroelec. Freq. Control* **2002**, *49*, 585–595. [CrossRef]

28.  Akasheh, F.; Fraser, J.D.; Bose, S.; Bandyopadhyay, A. Piezoelectric micromachined ultrasonic transducers: Modeling the influence of structural parameters on device performance. *IEEE Trans. Ultrason. Ferroelectr. Freq. Control* **2005**, *52*, 455–468. [CrossRef] [PubMed]

29.  Na, S.; Wong, L.L.P.; Chen, A.I.; Li, Z.; Macecek, M.; Yeow, J.T.W. A CMUT Array Based on Annular Cell Geometry for Air-coupled Applications. In Proceedings of the IEEE International Ultrasonic Symposium, Tours, France, 18–21 September 2016; pp. 1–4.

30.  Roh, Y.; Khuri-Yakub, B.T. Finite element analysis of underwater capacitor micromachined ultrasonic transducers. *IEEE Trans. Ultrason. Ferroelectr. Freq. Control* **2002**, *49*, 293–298. [CrossRef] [PubMed]

31.  Olçum, S. Optimization of the Gain-Bandwidth product of capacitive Micromachined Ultrasonic Transducers Circuit Modeling of a CMUT. *Analysis* **2006**, *52*, 1–18.

32.  Olcum, S.; Yamaner, F.Y.; Bozkurt, A.; Atalar, A. Deep-collapse operation of capacitive micromachined ultrasonic transducers. *IEEE Trans. Ultrason. Ferroelectr. Freq. Control* **2011**, *58*, 2475–2483. [CrossRef] [PubMed]

33.  Zhuang, X.; Wygant, I.O.; Yeh, D.T.; Nikoozadeh, A.; Oralkan, O.; Ergun, A.S.; Cheng, C.H.; Huang, Y.; Yaralioglu, G.G.; Khuri-Yakub, B.T. Two-Dimensional Capacitive Micromachined Ultrasonic Transducer (CMUT) Arrays for a Miniature Integrated Volumetric Ultrasonic Imaging System. In *Medical Imaging 2005: Ultrasonic Imaging and Signal Processing*; SPIE: Bellingham, WA, USA, 2005; Volume 5750, pp. 37–46.

34.  Zhuang, X.; Ergun, A.S.; Huang, Y.; Wygant, I.O.; Oralkan, O.; Khuri-yakub, B.T. Integration of trench-isolated through-wafer interconnects with 2d capacitive micromachined ultrasonic transducer arrays. *Sens. Actuators A Phys.* **2007**, *138*, 221–229. [CrossRef] [PubMed]

35.  Zhuang, X.; Ergun, A.S.; Oralkan, O.; Wygant, I.O.; Khuri-yakub, B.T. Interconnection And Packaging For 2d Capacitive Micromachined Ultrasonic Transducer Arrays Based On Through -Wafer Trench Isolation. In Proceedings of the 19th IEEE International Conference on Micro Electro Mechanical Systems, Istanbul, Turkey, 22–26 January 2006; pp. 270–273.

36.  Mukhiya, R.; Sinha, A.; Prabakar, K.; Raghuramaiah, M.; Jayapandian, J.; Gopal, R.; Khanna, V.K.; Shekhar, C. Fabrication of capacitive micromachined ultrasonic transducer arrays with isolation-trenches using anodic wafer bonding. *IEEE Sens. J.* **2015**, *15*, 5177–5184. [CrossRef]

37. Töpper, M.; Ndip, I.; Erxleben, R.; Brusberg, L.; Nissen, N.; Schröder, H.; Yamamoto, H.; Todt, G.; Reichl, H. 3-D thin film interposer based on TGV (Through Glass Vias): An alternative to Si-interposer. In Proceedings of the IEEE 60th Electronic Components and Technology Conference, Las Vegas, NV, USA, 1–4 January 2010; pp. 66–73.

38. Zhang, X.; Yamaner, F.Y.; Oralkan, Ö. Fabrication of Vacuum-Sealed Capacitive Micromachined Ultrasonic Transducers With Through-Glass-Via Interconnects Using Anodic Bonding. *J. Microelectromech. Syst.* **2017**, *26*, 226–234. [CrossRef]

39. Hu, X.; Bäuscher, M.; Mackowiak, P.; Zhang, Y.; Hoelck, O.; Walter, H.; Ihle, M.; Ziesche, S.; Hansen, U.; Maus, S.; et al. Characterization of Anodic Bondable LTCC for Wafer-Level Packaging. In Proceedings of the 2016 IEEE 18th Electronics Packaging Technology Conference (EPTC), Singapore, 30 November–3 December 2016; pp. 501–505.

40. Tanaka, S. Wafer-level hermetic MEMS packaging by anodic bonding and its reliability issues. *Microelectron. Reliab.* **2014**, *54*, 875–881. [CrossRef]

41. Tanaka, S.; Matsuzaki, S.; Mohri, M.; Okada, A.; Fukushi, H.; Esashi, M. Wafer-level hermetic packaging technology for MEMS using anodically-bondable LTCC wafer. In Proceedings of the 2011 IEEE 24th International Conference on Micro Electro Mechanical Systems, Cancun, Mexico, 23–27 January 2011; pp. 376–379.

42. Tanaka, S.; Mohri, M.; Ogashiwa, T.; Fukushi, H.; Tanaka, K.; Nakamura, D.; Nishimori, T.; Esashi, M. Electrical interconnection in anodic bonding of silicon wafer to LTCC wafer using highly compliant porous bumps made from submicron gold particles. *Sens. Actuators A Phys.* **2012**, *188*, 198–202. [CrossRef]

43. Mohri, M.; Esashi, M.; Tanaka, S. MEMS Wafer-Level Packaging Technology Using LTCC Wafer. *Electron. Commun. Jpn.* **2014**, *97*, 42–51. [CrossRef]

44. Yildiz, F.; Matsunaga, T.; Haga, Y. Capacitive micromachined ultrasonic transducer arrays incorporating anodically bondable low temperature co-fired ceramic for small diameter ultrasonic endoscope. *Micro Nano Lett.* **2016**, *11*, 627–631. [CrossRef]

45. Yildiz, F.; Matsunaga, T.; Haga, Y. CMUT Arrays Incorporating Anodically Bondable LTCC for Small Diameter Ultrasonic Endoscope. In Proceedings of the 11th IEEE Annual International Conference on Nano/Micro Engineered and Molecular Systems (NEMS), Sendai, Japan, 17–20 April 2016; pp. 50–53.

46. Yildiz, F.; Haga, Y.; Matsunaga, T. Capacitive Micromachined Ultrasonic Transducer Packaging for Forward-looking Ultrasonic Endoscope using Low Temperature Co-fired Ceramic Side Via. *IEEJ Trans. Sens. Micromach.* **2016**, *136*, 515–521. [CrossRef]

47. Tamaki, S.; Kuki, T.; Matsunaga, T. Flexible Tube-Shaped Neural Probe for Recording and Optical Stimulation of Neurons at Arbitrary Depths. *Sens. Mater.* **2015**, *27*, 507–523.

48. Matsunaga, T.; Matsuoka, Y.; Ichimura, S.; Wei, Q.; Kuroda, K.; Kato, Z.; Esashi, M.; Haga, Y. Multilayered receive coil produced using a non-planar photofabrication process for an intraluminal magnetic resonance imaging. *Sens. Actuators A Phys.* **2017**, *261*, 130–139. [CrossRef]

49. Bauerle, J.E. Study of Solid Electrolyte by a Complex Admittance Method. *J. Phys. Chem. Solids* **1969**, *30*, 2657–2670. [CrossRef]

50. Ahmad, B.; Pratap, R. Analytical evaluation of squeeze film forces in a CMUT with sealed air-filled cavity. *IEEE Sens. J.* **2011**, *11*, 2426–2431. [CrossRef]

51. Soni, S.; Jain, N.K.; Joshi, P.V. Vibration analysis of partially cracked plate submerged in fluid. *J. Sound Vib.* **2018**, *412*, 28–57. [CrossRef]

52. Walter, V.; Bourbon, G.; Le Moal, P. Residual stress in capacitive micromachined ultrasonic transducers fabricated with Anodic Bonding using SOI wafer. *Procedia Eng.* **2014**, *87*, 883–886. [CrossRef]

53. Si, X.H.; Lu, W.X.; Chu, F.L. Modal analysis of circular plates with radial side cracks and in contact with water on one side based on the RayleighRitz method. *J. Sound Vib.* **2012**, *331*, 231–251. [CrossRef]

*Article*

# A Mini-System Integrated with Metal-Oxide-Semiconductor Sensor and Micro-Packed Gas Chromatographic Column

**Jianhai Sun [1],\*, Zhaoxin Geng [2], Ning Xue [1],\*, Chunxiu Liu [1] and Tianjun Ma [1]**

[1]  State Key Laboratory of Transducer Technology, Institute of Electronics, Chinese Academy of Sciences, Beijing 100190, China; cxliu@mail.ie.ac.cn (C.L.); mmmmtj@126.com (T.M.)

[2]  School of Science, Minzu University of China, Beijing 100081, China; zxgeng@semi.ac.cn

\*  Correspondence: jhsun@mail.ie.ac.cn (J.S.); xuening@mail.ie.ac.cn (N.X.);
Tel.: +86-010-588-871-83 (J.S. & N.X.)

Received: 28 May 2018; Accepted: 7 August 2018; Published: 17 August 2018

**Abstract:** In this work, a mini monitoring system integrated with a microfabricated metal oxide array sensor and a micro packed gas chromatographic (GC) column was developed for monitoring environmental gases. The microfabricated packed GC column with a 1.6 m length was used to separate the environmental gas, and the metal oxide semiconductor (MOS) array sensor, fabricated with nano-sized $SnO$-$SnO_2$ sensitive materials, was able to effectively detect each component separated by GC column. The results demonstrate that the monitoring system can detect environmental gas with high precision.

**Keywords:** packed gas chromatographic column; metal-oxide-semiconductor array sensor; sensitive material; environmental monitoring

## 1. Introduction

With the development of industry, factories and automobiles are crowded in cities, making the city's environment full of harmful gases, such as CO, benzene, $SO_2$, and others, which are especially damaging to human health [1,2]. Moreover, the harmful components mixed in the air are very complicated. For the purpose of air filtering or air quality monitoring, these harmful components need to be accurately identified and monitored so that these harmful components can be effectively removed from the air or real-time detected. Therefore, the demand on gas sensors with high performance is very urgent.

MOS (metal oxide semiconductor) sensors which are surface-modified with different sensitive materials can detect different gases [3–19], such as carbon monoxide, sulfur dioxide, hydrogen sulfide, ammonia, and so on. Therefore, MOS sensors have become important environmental gas detectors and have attracted the attention of many researchers. In recent years, a large number of research papers on various MOS detectors have been reported. Nicoletti S and other research teams [4–15] have developed MOS gas detectors using nanocomposites as sensitive materials. Tomer V and other research teams [16,17] demonstrated MOS sensors with high performance that were doped with Ag or other catalytic materials. Kim [18] and other groups [19–24] also have successfully developed a variety of MOS sensors using sensitive materials based on binary or multiple metal oxides which have better reactivity towards target gases than single oxides. In addition, some research groups have proposed some other methods to improve selectivity and sensitivity of MOS sensors [25–28].

MOS sensors can detect environmental gas with high sensitivity, however, MOS sensors have poor resolution, and the problem of cross-talk interference between gases with similar properties is very serious. For example, when NO is detected, $NO_2$ will interfere with it. Similarly, $SO_2$ will also interfere with the detection of $H_2S$. Chromatography is a powerful analytical technique, which is able to separate mixed gases, as the retention time of gases in the stationary phase is different. Then the completely separated gases were quantified by high sensitive detectors. This method avoids the cross-talk interference between gases, thus greatly improving detection precision. In recent years, miniaturized gas chromatography columns (GCCs) have received increased attention and are under development in many laboratories [29–32].

Therefore, in this work, a microfabricated metal oxide array sensors based on nanosized $SnO$-$SnO_2$ sensitive materials were fabricated. Compared to pure $SnO_2$ sensitive materials, the $SnO_2$ sensitive material doped with $Sn^{2+}$ has more oxygen vacancies, which has a property of N type semiconductors and higher gas activity. Moreover, in order to solve the cross-talk interference between gases, in this work, a micro GC column, which has powerful separation ability, was proposed to separate analytical sample. After these interfering gas components were separated by the GC column, they were able to be detected with high precision by the MOS detector.

## 2. Materials and Methods

### 2.1. Materials and Reagents

In this work, in order to evaluate sensitivity of the fabricated MOS array sensor, carcinogenic gas, benzene (Sample I, purchased from Beijing Hua Yuan Gas Chemical Industry Co., Ltd., Beijing, China) was used as the test target; the concentration of benzene is 5 ppm. Sample II (Beijing Hua Yuan Gas Chemical Industry Co., Ltd., Beijing, China) was composed of 3 compounds (benzene, CO, and $SO_2$, the concentrations are 5 ppm, 500 ppm, and 505 ppm, respectively). Porapak Q with a diameter of 100 μm was purchased from Sigma-Aldrich (St. Louis, MO, USA).

### 2.2. Experimental Setup

In this work, the mini system integrated with a micro GC column and micro MOS array sensor was proposed for monitoring environmental harmful gas. The sample injected by a sampling pump was transported into the packed column through a valve purchased from Valco Instruments Company Incorporated (VICI). Clean air was used as carrier gas, the flow rate of which was controlled by electronic pressure control (EPC) technique. The setup of the system is shown in Figure 1. The working principle of the system is as follows. Firstly, a certain amount of sample was collected by sampling pump, and the sample was directly transported into the detector through port 1 of the solenoid valve, which measured the total amount of gas in the sample. Then, a certain amount of gas was collected and transported into the tube through port 2 of the solenoid valve when the diaphragm valve is in a closed state. Finally, the sampling probe was detached from the pollutant source, and fresh air, which acted as carrier gas, was pumped and transported through the sample into the micro GC column by opening the diaphragm valve. Then these components were separated by the micro GC column and detected by the MOS sensor. In this work, a micro GC column with length of 1.6 m [32] was used for separating polluted gases is a packed column developed by our lab. Channels of the packed GC column were fabricated using a laser etching technology (LET) which can easily fabricate deep well-shaped channels on glass wafer or silicon wafer, and the fabricated column (refer to Figure 2) with a rectangular cross section of 1.2 mm (depth) × 0.6 mm (width) has a large aspect ratio of 2:1. In order to effectively separate the polluted gas (such as CO, $SO_2$, and benzene, etc.), Porapak Q, acting as stationary phase, was packed in the column, and the packing process is detailed as follows. First, the inlet of the column was emerged into the porapak Q powder and the outlet was connected with a pump (the flow rate of the pump needs to be in the range of 1–5 L/min). Then, the porapak Q powder was transported into the micro channels under the pumping action. In order to uniformly pack the column, the column needs to be gently beaten during the packing process.

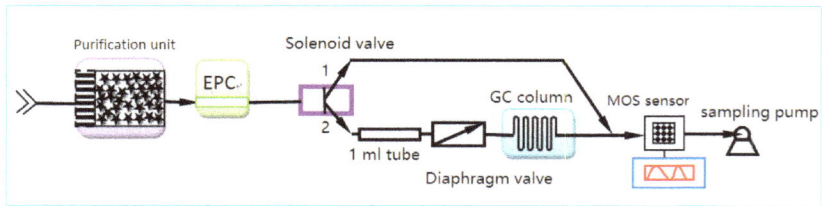

**Figure 1.** Configuration of the mini monitoring system.

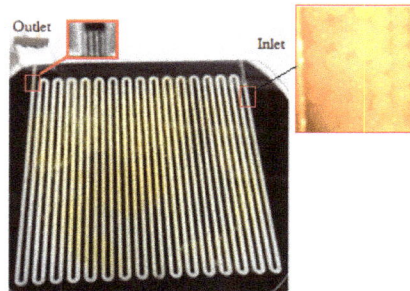

**Figure 2.** The fabricated micro GC column packed with porapak Q stationary phase.

*2.3. Fabrication of MOS Array Sensor*

It is well known that the MOS array sensor consist of a SnO-SnO$_2$ nano-metric film was deposited over a silicon micro-machined substrate implemented with platinum heater and two electrodes for contacting thin sensitive element. The MOS array sensor comprising of four completely independent detectors was fabricated based on micro-electro-mechanical system (MEMS) technology, and silicon was used as substrate wafers. The fabrication process of the chip was accomplished as follows. (1) A layer of boron ions (B$^+$) with thickness of 5 μm was implanted into the surface of silicon as a mask for the corrosion of silicon and a supported beam for hotplate and electrodes. The concentration of B$^+$ was $1 \times 10^{19}$ cm$^{-3}$, which was high enough for the self-stop corrosion. (2) A layer of SiN film with thickness of 300 nm was deposited on silicon wafer using low pressure chemical vapor deposition (LPCVD) technology. (3) The hotplate was fabricated as a 20 nm/150 nm Cr/Pt stack deposited by the magnetron sputtering technology and patterned by the lift-off technology. (4) A layer of SiN film with thickness of 200 nm acted as passivation layer (to provide electrical insulation between the platinum heater and the sensing layer) was deposited on the top of the Pt hotplate using plasma enhanced chemical vapor deposition (PECVD) technology. (5) The electrodes were realized as a 30 nm/200 nm Cr/Au stack deposited by the magnetron sputtering technology and patterned by the lift-off technology.

SnO$_2$ is a direct broadband gap semiconductor material. SnO$_2$ has a wide bandgap; a pure and ideal chemical ratio SnO$_2$ has high resistance. However, after the SnO$_2$ material was doped with SnO, the SnO$_2$ material deviates from its ideal chemical ratio, which makes the lattice with mixed normal ions (Sn$^{2+}$, Sn$^{4+}$) and oxygen negative ions (O$^{2-}$) in the absence of state, easily producing oxygen vacancy. As oxygen vacancies are able to form two donor levels in the forbidden band, and the two donor levels formed by oxygen vacancies have been completely dissociated at room temperature, SnO$_2$ has the property of N type semiconductors. The N type SnO$_2$ thin film has excellent performance. Its carrier concentration can reach $10^{19}$–$10^{21}$ cm$^{-3}$, and its conductivity can reach $10^{-3}$–$10^{-2}$ Ω·cm. Therefore, in this work, in order to improve the activity of SnO$_2$ sensitive material, SnO materials are doped into SnO$_2$ sensitive materials according to a certain mass ratio. The sensitive film was fabricated and the process was defined as follows, and a schematic representation of the whole structure is depicted in the Figure 3.

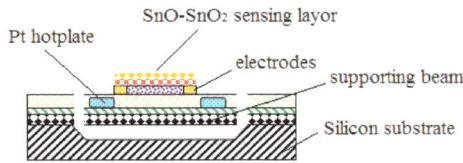

**Figure 3.** A schematic representation of the whole structure.

First, over the surface of Au electrodes (for contacting thin sensitive film), a layer of SnO and SnO$_2$ thin film deposited over the hotplate surface have been carried out by sputtering technology, and the thickness of the SnO and SnO$_2$ thin film were 50 nm and 150 nm, respectively. In order to increase the selectivity and sensitivity of the sensitive film, an extremely thin Au film acting as catalytic material with thickness of 5 nm was deposited over its surface. Noble metals [33] Au, as a surface active center, has the function of catalytic oxidation. In addition, the noble metal Au has large electronic affinity, which can accelerate the transfer of electrons from semiconductors to a noble metal and improve the sensitivity. Finally, the release process of the supporting beam was shown as follows. A layer of photoresist with thickness of 2 μm was coated and patterned as an etch mask for silicon nitride. After the two layers of silicon nitride were etched by reactive-ion etching (RIE) technology, a deep reactive-ion etching (DRIE) process was utilized to remove the diffusion of silicon in the micro channels. Then the supporting beam was released through a silicon etch (using 40% wt% KOH solution at 80 °C for 70 min). The width and length of chip (refer to Figure 4) was 8 mm and 10 mm, respectively, and the active area of the sensor (for each of the four sensors) was only 1 × 4 mm$^2$, consisting of a platinum resistor acting as heater.

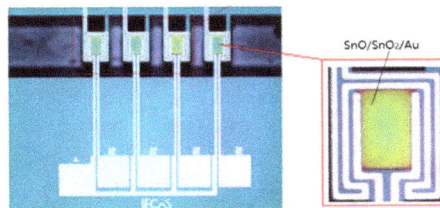

**Figure 4.** Photo of the MOS array sensor.

## 3. Results

### 3.1. Gas Sensing Characteristics of the MOS Array Sensors

In order to evaluate sensitivity of the fabricated MOS array sensor based on SnO-SnO$_2$ sensitive material, a MOS sensor coated with pure SnO$_2$ as a sensitive material was used for a comparison experiment. Table 1 shows the response characteristics of the two sensors to same samples (sample II). From the response amplitude of the output, the sensitivity of the MOS sensor doped with SnO is much greater than that of the MOS sensor coated with pure SnO$_2$ only.

**Table 1.** The average response of the two sensors to same sample.

| Average Response | MOS Sensor (Coated with SnO/SnO$_2$) | MOS Sensor (Coated with SnO$_2$ Only) |
|---|---|---|
| The average response of benzene (mV) | 290.0 | 165.0 |
| The average response of CO (mV) | 120.0 | 75.0 |
| The average response of SO$_2$ (mV) | 85.0 | 48.0 |

The concentration response characteristics of the MOS array sensor fabricated with a nano-sized SnO-SnO$_2$ sensitive material were analyzed and evaluated. Samples containing benzene were diluted to different concentrations and successively transported directly to the MOS sensor without passing through the micro GC column in periodically. Figure 5 shows the response curve of the fabricated sensor. As we can see that the detector can obtain an obvious response to extremely low concentrations of gas, moreover, the sensor has a response gradient to different concentration gases, and there is a linear relation between the increase of output signal and the concentration.

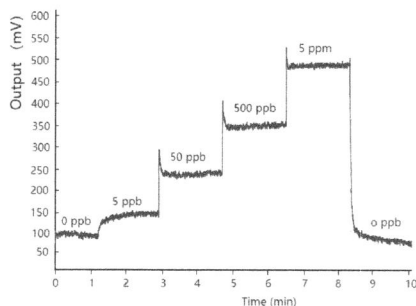

**Figure 5.** Response of the fabricated sensor to benzene at different concentrations.

### 3.2. Rapid Detection of Polluted Gases

To evaluate the performance of the fabricated monitoring system integrated with micro GC column and micro MOS array sensor. The experiment was carried out with sample II at a flow rate of 10 mL/min, and the inlet pressure of column was 80 psi and temperature of the column was 80 °C. As can be seen from the chromatogram (refer to Figure 6), the MOS array sensor integrated with chromatographic column was able to detect each component without mutual interference, solving the technical bottleneck of cross-talk interference between gases. That is to say, this detection method makes good use of the function of chromatography to solve the inherent technical bottleneck of MOS sensor. The specific concentration of each component can be accurately defined by comparing the area of each chromatographic peak with the corresponding peak area of the known concentration. Therefore, the high sensitive detector can overcome its own defects by integrating chromatography, which will greatly expand its applications and play an important role in environmental monitoring.

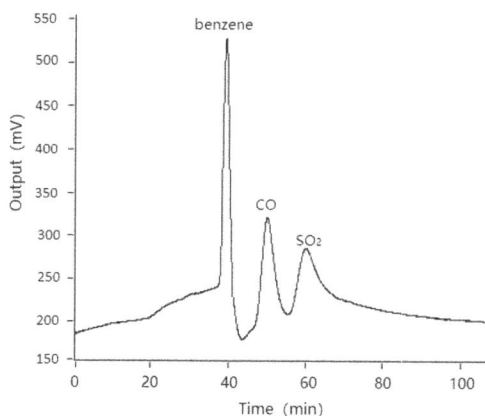

**Figure 6.** Chromatogram of benzene, CO, and SO$_2$.

*Micromachines* **2018**, *9*, 408

## 4. Conclusions

The work here demonstrated that it was possible to fabricate a mini monitoring system integrated with a MOS array sensor and a micro packed chromatographic column. By using a powerful chromatographic separation capability, the MOS array sensor was able to detect each component with high resolution and solve the technical bottleneck of mutual interference between gases. Therefore, the fabricated MOS array detector with high sensitivity can overcome its own defects by combining chromatographic techniques, which will greatly expand its applications and play an important role in environmental monitoring.

**Author Contributions:** Conceptualization, J.S. and N.X. Methodology, Z.G. Investigation, C.L. Writing, original draft preparation, J.S. and T.M.

**Funding:** This work has been partially financed by the National Key Research and Development Program of China Project (project ID: 2016YFC07006) and financed by the National Science Foundation of China Project (project ID: 61874121, 61774157, 61802363, 11574219, and 61176112).

**Conflicts of Interest:** The authors declare no conflict of interest.

## References

1. Pekkanen, J.; Peters, A.; Hoek, G.; Tiittanen, P.; Brunekreef, B.; de Hartog, J.; Heinrich, J.; Ibald-Mulli, A.; Kreyling, W.G.; Lanki, T.; et al. Particulate air pollution and risk of ST-segment depression during repeated submaximal exercise tests among subjects with coronary heart disease. *Circulation* **2002**, *106*, 933–938. [CrossRef] [PubMed]
2. Peters, A.; Dockery, D.W.; Muller, J.E.; Mittleman, M.A. Increased particulate air pollution and the triggering of myocardial infarction. *Circulation* **2001**, *103*, 2810–2815. [CrossRef] [PubMed]
3. Naik, A.; Parkin, I.; Binions, R. Gas sensing studies of an N-N hetero-junction array based on $SnO_2$ and ZnO composites. *Chemosensors* **2016**, *4*, 3. [CrossRef]
4. Nicoletti, S.; Zampolli, S.; Elmi, I.; Dori, L.; Severi, M. Use of different sensing materials and deposition techniques for thin-film sensors to increase sensitivity and selectivity. *IEEE Sens. J.* **2003**, *3*, 454–459. [CrossRef]
5. Elmi, I.; Zampolli, S.; Cardinali, G.C. Optimization of a wafer-level process for the fabrication of highly reproducible thin-film MOX sensors. *Sens. Actuators B Chem.* **2008**, *131*, 548–555. [CrossRef]
6. Sonker, R.K.; Sabhajeet, S.R.; Singh, S.; Yadav, B.C. Synthesis of ZnO nanopetals and its application as $NO_2$ gas sensor. *Mater. Lett.* **2015**, *152*, 189–191. [CrossRef]
7. Leonard, C.; Liu, H.F.; Brewer, S.; Sacks, R.D. High-speed gas extraction of volatile and semivolatile organic compounds from aqueous samples. *Anal. Chem.* **1998**, *70*, 3498–3504. [CrossRef]
8. Lee, D.S.; Jung, J.K.; Lim, J.W.; Huh, J.S.; Lee, D.D. Recognition of volatile organic compounds using $SnO_2$ sensors array and pattern recognition analysis. *Sens. Actuators B Chem.* **2001**, *77*, 228–236. [CrossRef]
9. Sberveglieri, G.; Faglia, G.; Groppelli, S.; Nelli, P.; Camanzi, A. A new technique for growing large surface area $SnO_2$ thin film (RGTO technique). *Semicond. Sci. Technol.* **1990**, *5*, 1231–1233. [CrossRef]
10. Liu, X.; Ma, T.; Pinna, N.; Zhang, J. Two-dimensional nanostructured materials for gas sensing. *Adv. Funct. Mater.* **2017**, *27*, 1702168. [CrossRef]
11. Zhang, J.; Liu, X.; Neri, G.; Pinna, N. Nanostructured materials for room-temperature gas sensors. *Adv. Mater.* **2016**, *28*, 795–831. [CrossRef] [PubMed]
12. Joshi, N.; Hayasaka, T.; Liu, Y.; Liu, H.; Oliveira, O.N.; Lin, L. A review on chemiresistive room temperature gas sensors based on metal oxide nanostructures, graphene and 2D transition metal dichalcogenides. *Microchim. Acta* **2018**, *185*, 213. [CrossRef] [PubMed]
13. Suman, P.H.; Felix, A.A.; Tuller, H.L.; Varela, J.A.; Orlandi, M.O. Comparative gas sensor response of $SnO_2$, SnO and $Sn_3O_4$ nanobelts to $NO_2$ and potential interferents. *Sens. Actuators B Chem.* **2015**, *208*, 122–127. [CrossRef]
14. Kim, J.H.; Zheng, Y.; Mirzaei, A.; Kim, H.W.; Kim, S.S. Synthesis and selective sensing properties of rGO/metal-coloaded $SnO_2$ nanofibers. *J. Electron. Mater.* **2017**, *46*, 3531–3541. [CrossRef]
15. Fan, H.; Xu, S.; Cao, X.; Liu, D.; Yin, Y.; Hao, H.; Wei, D.; Shen, Y. Ultra-long $Zn_2SnO_4$-ZnO microwires based gas sensor for hydrogen detection. *Appl. Surf. Sci.* **2017**, *400*, 440–445. [CrossRef]

16. Tomer, V.K.; Duhan, S. Ordered mesoporous Ag-doped $TiO_2/SnO_2$ nanocomposite based highly sensitive and selective VOC sensors. *J. Mater. Chem. A* **2016**, *4*, 1033–1043. [CrossRef]

17. Korotcenkov, G.; Brinzari, V.; Cho, B.K. Conductometric gas sensors based on metal oxides modified with gold nanoparticles: A review. *Microchim. Acta* **2016**, *183*, 1033–1054. [CrossRef]

18. Kim, J.H.; Lee, J.H.; Mirzaei, A.; Kim, H.W.; Kim, S.S. $SnO_2(n)$-NiO(p)composite nanowebs: Gas sensing properties and sensing mechanisms. *Sens. Actuators B Chem.* **2018**, *258*, 204–214. [CrossRef]

19. Zhou, Y.; Lin, X.; Wang, Y.; Liu, G.; Zhu, X. Study on gas sensing of reduced graphene oxide/ZnO thin film at room temperature. *Sens. Actuators B Chem.* **2017**, *240*, 870–880. [CrossRef]

20. Zhu, X.; Guo, Y.; Ren, H.; Gao, C.; Zhou, Y. Enhancing the $NO_2$ gas sensing properties of $rGO/SnO_2$ nanocomposite films by using microporous substrates. *Sens. Actuators B Chem.* **2017**, *248*, 560–570. [CrossRef]

21. Niu, X.S.; Du, W.P.; Du, W.M. Preparation and gas sensing properties of $ZnM_2O_4$ (M = Fe, Co, Cr). *Sens. Actuators B Chem.* **2004**, *99*, 405–409. [CrossRef]

22. Joshi, N.; da Silva, L.F.; Jadhav, H.S.; Shimizu, F.M.; Suman, P.H.; M'Peko, J.C.; Orlandi, M.O.; Seo, J.G.; Mastelaro, V.R.; Oliveira, O.N., Jr. Yolk-shelled $ZnCo_2O_4$ microspheres: Surface properties and gas sensing application. *Sens. Actuators B Chem.* **2018**, *257*, 906–915. [CrossRef]

23. Sun, F.J.; Li, X.G.; Liu, L.P.; Wang, J. Novel Zn–M–O (M = Sn, Co) sensing electrodes for selective mixed potential $CO/C_3H_8$ sensors. *Sens. Actuators B Chem.* **2013**, *184*, 220–227. [CrossRef]

24. Zhang, G.Y.; Guo, B.; Chen, J. $MCo_2O_4$ (M = Ni, Cu, Zn) nanotubes: Template synthesis and application in gas sensors. *Sens. Actuators B Chem.* **2006**, *114*, 402–409. [CrossRef]

25. Song, Z.; Liu, J.; Liu, Q.; Yu, H.; Zhang, W. Enhanced $H_2S$ gas sensing properties based on $SnO_2$ quantum wire/reduced graphene oxide nanocomposites: Equilibrium and kinetics modeling. *Sens. Actuators B Chem.* **2017**, *249*, 632–638. [CrossRef]

26. Huang, J.; Du, Y.; Wang, Q.; Zhang, H.; Geng, Y.; Li, X.; Tian, X. UV-enhanced ethanol sensing properties of RF magnetron-sputtered ZnO film. *Sensors* **2018**, *18*, 50. [CrossRef] [PubMed]

27. Wolfrum, E.J.; Meglen, R.M.; Peterson, D.; Sluiter, J. Metal oxide sensor arrays for the detection, differentiation, and quantification of volatile organic compounds at sub-parts-per-million concentration levels. *Sens. Actuators B Chem.* **2006**, *115*, 322–329. [CrossRef]

28. Zhang, D.; Liu, J.; Jiang, C.; Liu, A.; Xia, B. Quantitative detection of formaldehyde and ammonia gas via metal oxide-modified graphene-based sensor array combining with neural network model. *Sens. Actuators B Chem.* **2017**, *240*, 55–65. [CrossRef]

29. Reddy, K.; Jing, L.; Oo, M.; Fan, X. Integrated separation columns and Fabry-Perot sensors for micro gas chromatography systems. *IEEE J. Microelectromech. Syst.* **2013**, *22*, 1174–1179. [CrossRef]

30. Sun, J.H.; Guan, F.Y.; Cui, D.F.; Chen, X.; Zhang, L.L. An improved photoionization detector with a micro gas chromatography column for portable rapid gas chromatography system. *Sens. Actuators B Chem.* **2013**, *188*, 513–518. [CrossRef]

31. Ali, S.; Ashraf-Khorassani, M.; Taylor, L.T.; Agah, M. MEMS-based semi-packed gas chromatography columns. *Sens. Actuators B Chem.* **2009**, *14*, 309–315. [CrossRef]

32. Sun, J.H.; Guan, F.Y.; Zhu, X.F.; Ning, Z.W.; Ma, T.J.; Liu, J.H.; Deng, T. Micro-fabricated packed gas chromatography column based on laser etching technology. *J. Chromatogr. A* **2016**, *1429*, 311–316. [CrossRef] [PubMed]

33. Kohl, D. The role of noble metals in the chemistry of solid state gas sensors. *Sens. Actuators B. Chem.* **1990**, *21*, 158–165. [CrossRef]

*micromachines*

MDPI

*Article*

# The Fabrication of Au@C Core/Shell Nanoparticles by Laser Ablation in Solutions and Their Enhancements to a Gas Sensor

**Xiaoxia Xu [†], Lei Gao [†] and Guotao Duan ***

Key Laboratory of Materials Physics, Anhui Key Laboratory of Nanomaterials and Nanotechnology,
Institute of Solid State Physics, Chinese Academy of Sciences, Hefei 230031, China; xxxu@issp.ac.cn (X.X.);
lgao@issp.ac.cn (L.G.)
* Correspondence: duangt@issp.ac.cn
† These authors contributed equally to this work.

Received: 16 March 2018; Accepted: 29 May 2018; Published: 1 June 2018

**Abstract:** A convenient and flexible route is presented to fabricate gold noble metal nanoparticles wrapped with a controllable ultrathin carbon layer (Au@C) in one step based on laser ablation of the noble metal targets in toluene-ethanol mixed solutions. The obtained metal nanoparticles were <20 nm in size after ablation, and the thickness of the wrapped ultrathin carbon layer was 2 nm in a typical reaction. The size of the inner noble metal nanoparticles could be controlled by adjusting the power of laser ablation, and the thickness of the ultrathin carbon layer can be controlled from 0.6 to 2 nm by laser ablation in different components of organic solution. Then the resultant Au@C core/shell nanoparticles were modified on the surface of $In_2O_3$ films through a sol-gel technique, and the hydrogen sulfide ($H_2S$) gas-sensing characteristics of the products were examined. Compared to pure and Au-modified $In_2O_3$, the Au@C-modified $In_2O_3$ materials exhibited a revertible and reproducible performance with good sensitivity and very low response times (few seconds) for $H_2S$ gas with a concentrations of 1 to 5 ppm at room temperature. Evidence proved that the ultrathin carbon layer played an important role in the improved $H_2S$ sensor performance. Other noble metals wrapped by the homogeneous carbon shell, such as Ag@C, could also be prepared with this method.

**Keywords:** laser ablation; core/shell nanostructure; ultrathin carbon layer; gas sensing

## 1. Introduction

Core/shell nanostructure nanocomposites, in which the inner nanoparticle is encapsulated and protected against agglomeration, adsorption, or chemical reaction by an outer shell, have attracted much attention due to their fantastic physical, chemical, biological, and catalytic properties [1–3]. The thickness of the outer shell plays a significant role in the performance of the nanocomposites. For example, the surface enhanced Raman scattering (SERS) of noble metal nanomaterials coated with special materials shell to detect some target molecules need the outer shell in nanometer scale, and the SERS signal will be weakened, or disappear, if the coating shell is too thick [4]. On the other hand, graphene or graphene-like structure materials process unique chemical and physical properties and have become a hot research topic since they appeared. As a sheath coating on nanoparticles they could effectively enhance the properties of the nanoparticles [5], and it requires the thickness of the outer carbon shell with one or several layers. Now there are even many methods to coat the nanoparticles, such as vapor deposition [6–8], solution dipping [9,10], sol-gel coating, and so on [11]. However, it is still a challenge to obtain the nanoparticles wrapped with ultrathin and homogeneous outer layers with the above methods, especially for the noble metal nanoparticles.

Laser ablation of the metal targets in liquids could easily and effectively prepare such structure materials in only one step [12–14]. Compared with the conventional method, it is an attractive green and versatile technique to prepare various metal nanoparticles, such as Au, Ag, Pt, and Ni, etc. The metal nanoparticles obtained by this method have excellent chemistry, metastable composition, easily functionalizable surfaces, high purity, and good dispersity, etc., and these properties are closely related to the applications such as catalysis and SERS research [13,15]. For example, using 532 nm output from a pulsed Nd:YAG laser (10 ns, 10 Hz) Shaji et al. successfully obtained ZnO nanoparticles with zinc metal as the target in distilled water at different water temperatures, and found that the morphology, structure, chemical state and optical properties of ZnO nanoparticles were closely related with the temperature and laser fluence [16]. Kautek et al. prepared $Ni/NiO_x$ core/shell nanoparticles in water and alcoholic fluids, and the nature of the fluid, the laser fluence, and the number of laser pulses decided the size distribution of the products. Through the changes of these parameters, the size distribution of $Ni/NiO_x$ core/shell nanoparticles was changed from 10 to 30 nm [17]. If only organic solvents are chosen as the liquids, it will be an ideal carbon source to form ultrathin carbon layers. By two sequential processes during ablation, i.e., formation of the noble metal nanoparticles and subsequent carbon-deposition, the homogeneous and ultrathin carbon layer-wrapped noble metal nanoparticles will be obtained in one step with this method. The production with a unique structure will be an effective and excellent modifying material for application of a semiconductor metal oxide-based gas sensor.

Semiconductor metal oxide-based gas sensors have received extensive research because they can detect the toxic, harmful, inflammable, or explosive gases quickly, efficiently, and accurately [18–21]. Among varieties of sensitive materials, indium oxide ($In_2O_3$) is a promising material for semiconductor gas sensor due to its peculiar properties, such as a wide band gap (3.56 eV), low resistance, and good catalysis [22–24]. Up to now, $In_2O_3$ has been widely applied to detect $H_2$, CO, $O_3$, and volatile organic compounds, etc. It is well known that the gas-sensing mechanism of semiconductors is based on the oxidation-reduction reaction between the surface of the sensor material and the test gas [25,26]. In order to improve the gas response and selectivity of semiconductor oxides, a nanoparticle-modifying method is often adopted, especially noble metal nanoparticles [27]. The reason is that it can easily change the electronic structure or space charge layer thickness of sensing films and improve the gas sensing performance. Herein, we report that modifying Au and Au@C nanoparticles prepared by laser ablation in liquid onto $In_2O_3$ film can lead to a greatly enhanced sensing sensitivity to $H_2S$ at room temperature.

## 2. Experimental Section

### 2.1. Au and Au@C Colloidal Solution Preparation

Typically, a gold (or silver) plate (99.99%, $1.5 \times 1$ cm$^2$, purchased from Sinopharm Chemical Reagent Limited Corporation, Shanghai, China) is fixed on a bracket in a quartz glass vessel and immersed in 10 mL toluene-ethanol mixed solution with a volume ratio of 9:1. The vessel was placed on a horizontal platform. The plate in the solution was irradiated, while the solution was continuously stirred, by the first harmonic of a Nd:YAG (yttrium aluminum garnet) laser (1064 nm in wavelength, 10 Hz in frequency, and 10 ns in pulse duration) with the power of 60 mJ/pulse and the spot size of about 2 mm in diameter on the plate. A purple colloidal solution was formed during the irradiation for 20 min. The ablation of the target in water was also carried out for reference. After ablation, the solution was centrifuged at 12,000 rpm, and the obtained Au@C colloidal solution was rinsed several times with ethanol in order to remove the residual toluene and dehydrated at 60 °C for 12 h. The Au colloidal solution prepared in water was centrifuged under the same condition and Au nanoparticles were dispersed in 2 mL ethanol.

### 2.2. Preparation of Au or Au@C Modified $In_2O_3$ Film-Based Sensing Devices

Firstly, 1.0 g indium nitrate hydrate powder was dispersed in 80 mL deionized water with continual stirring by a magnetic rotor for 10 min. Subsequently, 0.6 g urea was added to the above colorless precursor solution and stirred for 20 min to obtain a mixture solution. Secondly, a 2D ordered polystyrene spheres (PS) colloidal monolayer template on a glass slide with the sphere diameter of 1000 nm was prepared by air/water interfacial assembly. Then the PS monolayer colloidal template on the glass slide was peeled off and floated onto the surface of pure water in a beaker due to surface tension of the solution. A ceramic tube was used to pick up the PS monolayer colloidal template from the bottom of liquid surface and dried at 80 °C for 2 h. Thirdly, the ceramic tube coated by the PS monolayer colloidal template was impregnated to the above mixture solution and transferred into a sealed stainless steel reactor and kept at 90 °C for 2 h. After the reaction the ceramic tube was washed by deionized water three times to remove the excess mixture solution. Fourthly, the ceramic tube was heated at 300 °C for 2 h to burn the PS template away and the $In_2O_3$ thin film was formed on the tube. Finally, Au and Au@C nanoparticles were modified onto the $In_2O_3$ thin film by impregnating it in the Au and Au@C colloidal ethanol solution and dried at 60 °C for 1 h. Thus, the Au- or Au@C-modified $In_2O_3$ film-based sensing devices were fabricated.

*2.3. The Gas Sensing Measurements and Characterization Methods*

All the gas sensing performances were measured in a static system with a volume of 20 L by the measurement of electric circuit at room temperature, in which a fixed resistor with the values ranging from 0.1 to 100 M ohm were connected in series on a circuit board to adjust the voltage of the sensor. This system was consisted by several sections such as manometers, electromagnetic cut-off valves, hydrometer, measuring chamber, gas pressure stabilizers and the signal processing system. Agilent U8002A DC power (San Jose, CA, USA) provided a 10 V regulated power supply, and Agilent mod. U3606A (San Jose, CA, USA) collected and recorded the voltage change on the fixed resistor by the computer. The $H_2S$ gas with required quantities was injected into the measure system by a syringe, which was calculated by the ideal gas equation of state from the pristine gas cylinders of 10,000 ppm obtained commercially. Firstly the pure air was introduced into the measure system for 30 min and guaranteed the sensor's ambience with 60% RH, then a concentration of diluted $H_2S$ gas was injected and kept for 30 s, and the corresponding electric response signal of gas sensor was recorded by computer. Then the pure air was introduced into the measure system again to remove the $H_2S$ gas and after 60 s another concentration of diluted $H_2S$ gas was injected and kept for 30 s again, then cycled the above operation. In order to evaluate the selectivity performances of the sensor, some oxidizing or reducing gases such as $C_3H_6O$, $H_2$, $C_2H_6O$, $NH_3$, $CH_4$, and $SO_2$ were injected into the measurement system. A vaporizer was used to control the environmental humidity in the system, and a lab-made software was used to control the heating voltage and record data.

The resultant products were characterized by a field emission scanning electron microscope (FE-SEM, Hitachi, SU8020, Tokyo, Japan) and a transmission electron microscope (TEM, JEM-200CX, Tokyo, Japan). The optical absorption spectra were measured on a Cary-5E UV–VIS–NIR spectrophotometer (San Jose, CA, USA). The Raman spectra were recorded on a microscopic confocal Raman spectrometer (Renishaw inVia Reflex, London, UK) using a laser beam of 532 nm wavelength, 1 mW power and 5 μm spot size on the sample area. The gravimetric analysis was measured on TGA Q5000IR (Cranston, RI, USA).

## 3. Results and Discussion

In a typical reaction, using a toluene-ethanol mixed solution with a volume ratio of 9:1 as the laser ablation liquid, Au nanoparticles wrapped with an ultrathin carbon layer (Au@C) are obtained. The morphology and structure of the products are measured by TEM, as shown in Figure 1. It can be seen that the inner Au nanoparticles are spherical in structure with a diameter in the range of 5 to 15 nm and a mean size of 12 nm. Figure 1b is a larger image of Figure 1a, and the particles are much larger than the ones shown in Figure 1a. From the high-resolution TEM image of a partial particle, the clear crystal lattice can be discovered with a spacing of the lattice fringe of 0.24 nm, corresponding to the (111) plane of Au. Such a higher crystal surface index can ensure the prepared material with a better catalytic activity. The outer ultrathin carbon layer is about 2 nm in thickness with a spacing of the lattice fringe of 0.34 nm, which corresponds to the (001) plane of carbon. In addition, the energy spectrum of Au@C nanoparticles is also shown in the Figure S1, and it can be seen that the Au nanoparticles are indeed coated by a carbon layer, which corresponds to the results of high-resolution transmission electron microscopy (HRTEM). At the same conditions, using water solution as laser ablation liquids only Au nanoparticles with a size below

20 nm are formed after laser ablation of the Au target, and no carbon shell appears, as shown in Figure S2. Using water solution as laser ablation liquids only Ag nanoparticles without carbon shell is prepared with a size of about 50 nm after laser ablation of Ag target shown in Figure S3. Similarly, ablation of Ag target in the toluene-ethanol mixed solution with the same volume ratio leads to the wrapped Ag nanoparticles (Ag@C) with the average size of 15 nm, as shown in Figure S4. In addition, the energy spectrum of Ag@C nanoparticles is also shown in the Figure S5, which indicates that the Ag@C samples have been prepared successfully by the same method. The coating layer is also about 2 nm in thickness. In order to assess the amount of carbon in the shells of Au@C nanoparticles, a gravimetric analysis is measured and the results can be seen in the Figure S6. It can be seen that there is a mass loss of 0.94% at first, which corresponds to the small adsorbed molecules in the samples such as the water molecules. Then the other mass loss occurs from 199.7 °C, which corresponds to the combustion of the carbon shells. This indicates that the amount of carbon in the shells of Au@C nanoparticles is about 3.96%. The selected area electronic diffraction of Ag@C illustrated that the inner Ag nanoparticles have a good crystallinity. With this method, other types of metal nanoparticles wrapped with ultrathin carbon layers may also be obtained.

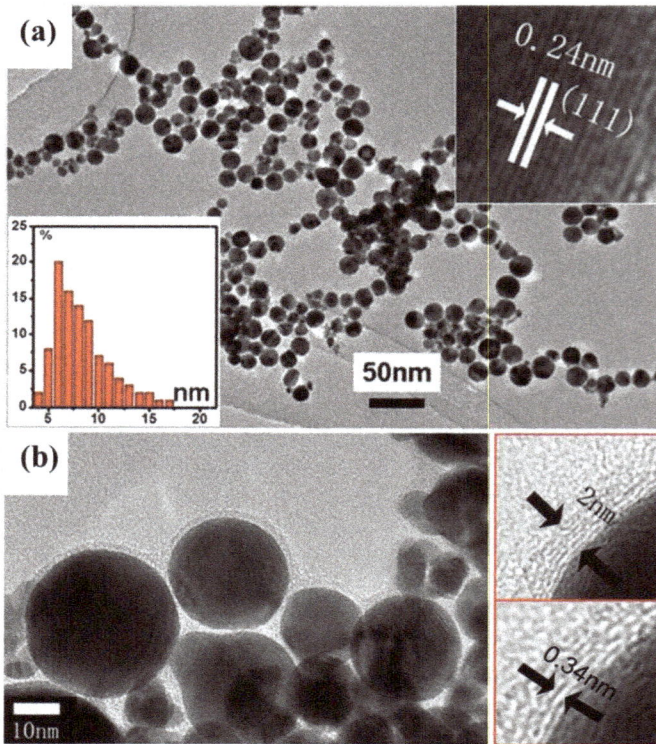

**Figure 1.** The TEM examination of the products induced by laser ablation of Au target in toluene-ethanol solution with the volume ratio 9:1 and the power of 60 mJ/pulse. (**a**): TEM image at low magnification. Insets: particle size distribution (lower-left) and high resolution TEM image of a partial particle showing Au (111) plane fringe (upper-right); and (**b**) local magnified image of (**a**). The insets show the thickness of the ultrathin carbon layer (upper-right) and the fringe spacing in the shell layer (lower-right).

The optical absorbance spectra of Au and Au@C colloidal solutions are shown in Figure 2a. In the measurement of optical absorbance spectra of all the samples, which are prepared at the same conditions with the power of 60 mJ/pulse and the spot size of about 2 mm in diameter on the plate for 20 min, 3 mL sample water solution (0.1 g/L) is put into a quartz cell with $12.5 \times 12.5 \times 45$ mm$^3$ and the same quartz cell with 3 mL pure water solution is used as a reference. Then they are measured with a wavelength of 300 to 800 nm. For ablation of the Au target in water, there exists an absorption band around 520 nm, which corresponds to the well-known surface plasmon resonance (SPR) of Au nanoparticles, indicating formation of Au nanoparticles in the water [28–31]. However, ablation in the toluene-ethanol solution only leads to a weak and broad absorption band around 550 nm, as shown in Figure 2a. Similarly, for Ag, there is a strong SPR of Ag nanoparticles around 400 nm after ablation in water, and a very small and broad absorption band around 415 nm for ablation in the mixed solution, as illustrated in Figure 2b. Obviously, ablation in the mixed solution induces the red-shift and significant decrease of the optical absorption band for the Au@C and Ag@C samples.

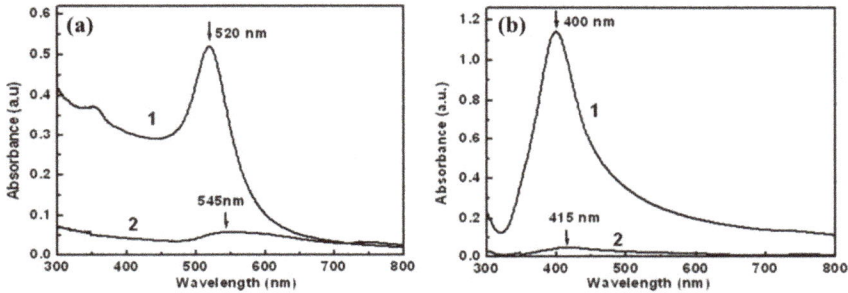

**Figure 2.** Optical absorbance spectra of the colloidal solutions induced by ablation of (**a**) Au target and (**b**) Ag target in solutions. Curve (1): ablation in water; and curve (2): ablation in the toluene-ethanol mixed solution with the volume ratio 9:1.

In order to further reveal the structural information about the carbon coating layer surrounding the Au nanoparticles, Raman spectral measurement is conducted for the carbon wrapped Au nanoparticles, and the result is shown in Figure 3a. It can be found that there are two broad peaks around 1350 cm$^{-1}$ and 1570 cm$^{-1}$, correspond to D band (1355 cm$^{-1}$) and G band (1590 cm$^{-1}$) of graphitic carbon, respectively [32,33]. For the carbon-wrapped Ag nanoparticles, the Raman spectrum is also similar, as shown in Figure 3b. Obviously, it can indicate that the outer shell is ultrathin carbon layer, corresponding to the results of TEM. Thus, the noble metal nanoparticles wrapped with controllable ultrathin carbon layer are fabricated by one-step based on laser ablation in ethanol-toluene mixed solutions.

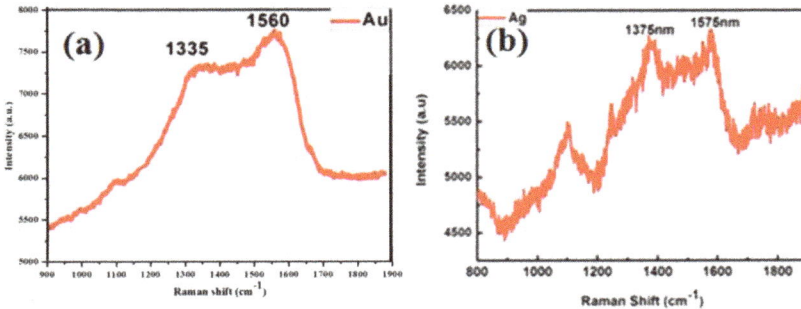

**Figure 3.** Raman spectrum for (**a**) Au@C colloidal solutions; and (**b**) Ag@C colloidal solutions.

In addition, the size of the inner noble metal nanoparticles can be controlled by the ablation power. Taking the Au@C nanoparticles as an example, the nanoparticles decrease from about 20 nm to about 10 nm in mean size with the decrease of the laser power from 100 to 40 mJ/pulse. The other conditions are unchanged, and the results can be seen in Figure 4. However, the thickness of the outer wrapping carbon layer is almost unchanged (~2 nm) at different ablation powers in our case. The reasons are that the higher power laser beams induce a higher density Au plasma plume on the target surface, and higher density Au particles are more easily nucleated to form larger size nanoparticles [16]. Therefore, the size of the inner noble metal nanoparticles can be controlled by adjusting the power of laser ablation, and the controllable growth of nanoparticles can be achieved by this method.

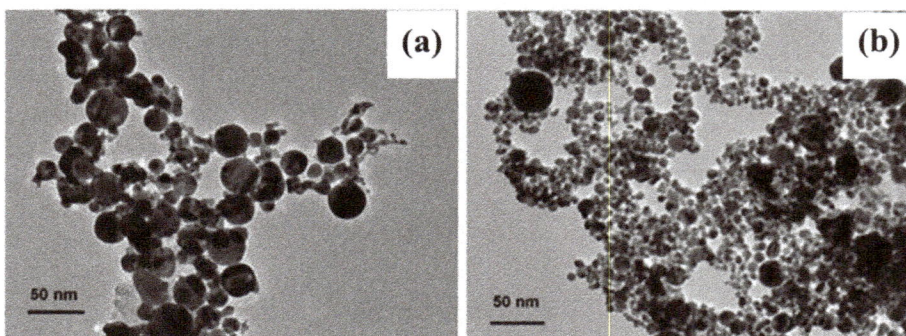

**Figure 4.** The TEM morphology characterization of the Au@C nanoparticles induced by the different ablation power. (**a**) 100 mJ/pulse; and (**b**) 40 mJ/pulse.

As expected, further experiments have revealed that the amount of the nanoparticles in the solution increases with the duration of the laser ablation from 2 to 60 min, but the thickness of the carbon shell is almost unchanged. However, the composition of liquid medium strongly influences the thickness of the outer carbon shell surrounding the inner noble metal nanoparticles, as shown in Figure 5. The higher content of the carbon component leads to the thicker carbon shell, or vice versa. In the water solution, there is no carbon shell generated, as shown in Figure 5a. The D band and G band of graphitic carbon also cannot be seen in Figure 5b. If laser ablates in the pure ethanol under the same conditions (60 mJ/pulse) as above, we can also obtain the Au@C nanoparticles but with much smaller thickness (only ~0.6 nm), as shown in Figure 5c. With the increase of the toluene content in the mixed solution, the carbon shell will get thicker and thicker. With the volume ratio of 1:1, the thickness of the outer carbon shell can reach to ~1.3 nm, as shown in Figure 5e, and ~2 nm with the volume ratio of 9:1 shown in Figure 2. The Raman spectral measurements have confirmed existence of carbon shell on the surface of noble metal nanoparticles (similar to that shown in Figure 3). Therefore, by laser ablation in different components of organic solution, we obtained different thickness of the ultrathin carbon layer, and successfully realize the effective control of the thickness of the outer carbon shell of the noble metal nanoparticles.

In addition, as shown in Figure S7, the color of the colloid solution is changed obviously in the four solutions of water, pure ethanol, toluene-ethanol mixed solution with volume ratios of 1:1 and 9:1, respectively. The color of the colloid solution of Au nanoparticles prepared in water is more transparent and red in color (A). The color is deep purple red for the product obtained in pure ethanol (B). In the toluene-ethanol mixed solution, the color is a darker purple-brown (C,D).

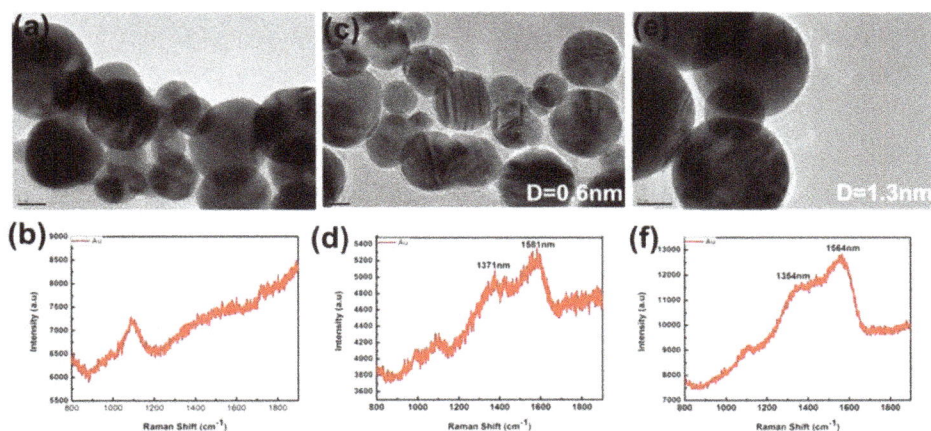

**Figure 5.** The TEM morphology characterization and Raman spectrum of the Au@C nanoparticles in different solutions. (**a,b**) water; (**c,d**) pure ethanol; and (**e,f**) toluene-ethanol mixed solution with the volume ratio of 1:1. The scales in (a), (e) and (f) are all 10 nm.

From the above results, the formation mechanism of the carbon-wrapped noble metal nanoparticles could be easily speculated as follows. When the laser beam irradiates on the surface of metal target, the high-pressure metal plasma will be quickly formed on the solid-liquid interface [34–36]. Subsequently, such metal plasma will ultrasonically and adiabatically expand, leading to cooling of the metal plume region and hence to formation of metal nanoparticles. At the same time, C-O and C-C ligands in ethanol or toluene molecules, at the interface between the metal plasma plume and the mixed solution, will be broken due to the extreme conditions to form carbon atoms. These carbon atoms would deposit on the preformed Au or Ag nanoparticles to form ultrathin carbon layer. For the laser ablation duration and power, it could be attributed to the number balance between the laser ablation-induced C and metal nanoparticles. The longer ablation duration or higher power will not only produce the more C-C broken bonds in the solvent molecules but also form the more Au or Ag nanoparticles. On the other hand, the thickness of the carbon shell surrounding the metal nanoparticles should significantly depend on the ability of liquid phase to supply carbon atoms at a certain laser fluence. Thus, some carbon-abundant solutions (such as toluene) could produce a thicker carbon layer than ethanol.

The $In_2O_3$ film is modified by Au@C nanoparticles (marked as $In_2O_3$/Au@C) with the solution impregnation method, which the Au@C nanoparticles are prepared in toluene-ethanol mixed solution with the volume ratio of 9:1 and the power of 60 mJ/pulse. From the Figure 6a, it can be seen that the $In_2O_3$ film on the ceramic tube is a dense granular film, and the main diffraction peaks of prepared $In_2O_3$ film is corresponding to cubic phase structure of $In_2O_3$ (JCPDS 74-1990) from its XRD pattern (Figure 6b). The TEM picture of the $In_2O_3$/Au@C sample and its energy spectrum can be seen in Figure 6c,d. On the surface of $In_2O_3$ film, there are many of spherical nanoparticles, which are the modifying Au@C nanoparticles. On the other hand, from the spectral peaks of Au elements, it is also proved that the Au@C nanoparticles are successfully modified on the surface of $In_2O_3$ granular film with solution impregnation method. The Au nanoparticles prepared by laser ablation in water solution are also modified on the $In_2O_3$ film (marked as $In_2O_3$/Au) with the same method.

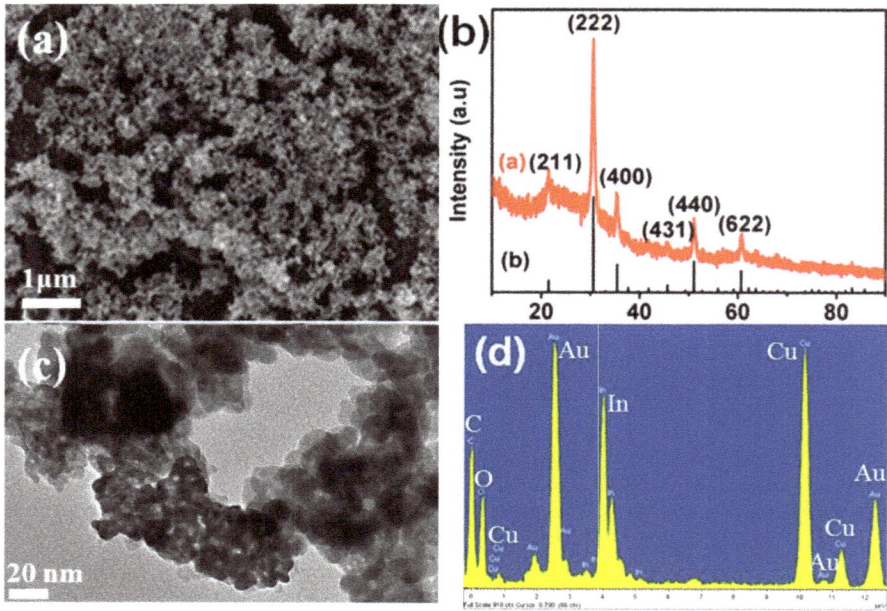

**Figure 6.** (**a**) SEM image of $In_2O_3$ thin film on ceramic tube; (**b**) corresponding XRD pattern; (**c**) TEM picture of as-prepared Au@C modified $In_2O_3$; and (**d**) its energy spectrum.

The gas-sensing properties of $In_2O_3$, $In_2O_3/Au$ and $In_2O_3/Au@C$ films to $H_2S$ gas are measured at room temperature from 1 to 5 ppm with a relative humidity (RH) of 60%, and the results can be seen in Figure 7 and Figure S8. For the $In_2O_3$ film at a concentration of 1 ppm, it has an excellent gas-sensitive response to $H_2S$ gas (about 90) at room temperature, but it cannot be recovered, as shown in Figure S8. It can be seen that for the $In_2O_3/Au@C$ films the sensing sensitivity is 52, 97, 141, 178, and 228 from the concentration of 1 to 5 ppm to $H_2S$ with RH = 60% at room temperature, respectively. For the $In_2O_3/Au$ films that is 24, 35, 47, 62, and 78 at the same conditions, respectively. Thus, the $In_2O_3/Au@C$ films have a better gas sensing sensitivity to $H_2S$ gas than $In_2O_3/Au$ films. However, the $In_2O_3/Au$ films have a better response and recovery time (defined as the times to reach 90% of resistance change) shown in Figure 7a. For example, at the concentration of 4 ppm to $H_2S$ gas, the response and recovery time is 9 and 20 s for the $In_2O_3/Au$ films, which is 16 and 33 s for the $In_2O_3/Au@C$ films. That may be related to the catalytic ability of Au nanoparticles, and for the $In_2O_3/Au$ films more Au nanoparticles are easily exposed to $H_2S$ gas, which are not coated by carbon layers. They will have a better catalytic performance and response and recovery capability. Additionally, for the $In_2O_3/Au@C$ films at the concentration of 5 ppm to $H_2S$ gas there is a sharp response curve, but it has achieved sensor signal saturation, which can be seen in the later discussion (the reproducibility response curves of $In_2O_3/Au@C$ sensor to $H_2S$ with a concentration of 5 ppm at room temperature in Figure 8). Similar to most semiconductor material, $In_2O_3$ is an n-type semiconducting metal oxide. When it is exposed to air, oxygen would be adsorbed on the surface of the $In_2O_3$ film and turned into chemisorbed oxygen, such as $O_2^{2-}$ or $O_2^-$, which plays a role as trap electrons and surface acceptors, and the resistance of the $In_2O_3$ film increases. If the $In_2O_3$ film is exposed to $H_2S$ gas, which is a strong reducing gas, the $H_2S$ molecules will react with the $O_2^{2-}$ or $O_2^-$ adsorbed on the surface of $In_2O_3$ film. Then the captured electrons will release back to the bulk, and the resistance of the $In_2O_3$ film decrease. Thus, the response of the samples to $H_2S$ gas $R_{air}/R_g$ will drastically increase. When the $In_2O_3$ film once is exposed to air, it will return to the initial state, and so on and so on. Exposed to high concentration of H$_2$S gas, the decrease of resistance of the In$_2$O$_3$ film is more obvious. The Au nanoparticles and carbon shell also plays an important role in the process, and the affect mechanism will discussed in detail later. For the two films, there is also a good linear relationship with concentration of H$_2$S gas as shown in Figure 7b, which is favorable to the practical application. For each concentration of H$_2$S gas, the number of repeat measurements is six and the resulting standard deviation for the two films can also be seen in Figure 7b.

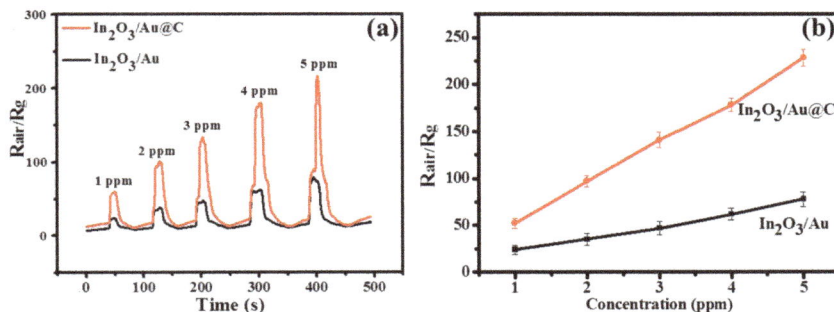

**Figure 7.** (**a**) The response curves as functions of test time to H$_2$S and (**b**) the sensitivity versus H$_2$S concentration of In$_2$O$_3$/Au@C and In$_2$O$_3$/Au sensor at a concentration ranging from 1–5 ppm with RH = 60% at room temperature.

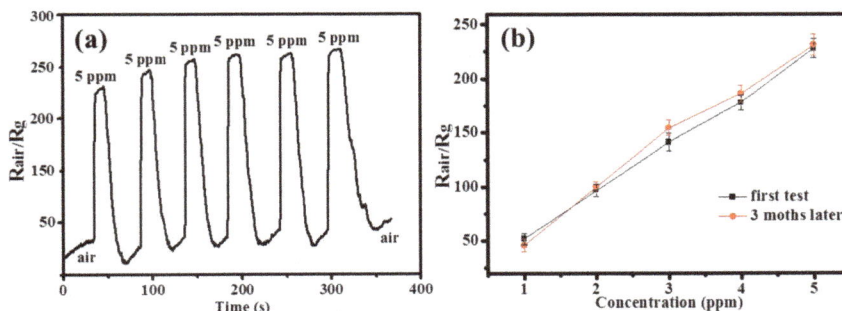

**Figure 8.** (**a**) The reproducibility response curves of In$_2$O$_3$/Au@C sensor to H$_2$S with a concentration of 5 ppm at room temperature; and (**b**) the concentration-dependent response curve to H$_2$S from 1–5 ppm at room temperature measured before and after three months for In$_2$O$_3$/Au@C sensor. All are tested under the ambience with 60% RH.

From the above results, it can be seen that the In$_2$O$_3$ films modified by Au@C have the best sensitivity. Thus, the practical performances of the In$_2$O$_3$/Au@C sensor are tested. In Figure 8a, it can be seen that the In$_2$O$_3$/Au@C gas sensor still have a similar response curves in the six cycles, and the sensing sensitivity is 237, 246, 252, 257, 258, and 261 with 60% RH to H$_2$S gas with a concentration of 5 ppm at room temperature, respectively. As the number of cycles increases, the sensing sensitivity for the H$_2$S gas increases as well. The reason is that at room temperature when the sample is exposed to H$_2$S gas with a concentration of 5 ppm, it will lead to the decrease of resistance of the In$_2$O$_3$/Au@C sample and the dramatic increase of the test signal (R$_{air}$/R$_g$). When it is exposed to air, the sample will return to the initial state and the test signal will decrease. However, in this process the ultrathin carbon shell may adsorb some residual H$_2$S gas, which results in the resistance of the In$_2$O$_3$/Au@C sample increase in air compared to the first state. As the number of cycles increases, more residual H$_2$S gas may be adsorbed on the surface of ultrathin carbon shell, and the test signal also drift higher.

Further, the concentration dependent response curve to H$_2$S from 1–5 ppm at room temperature with 60% RH is measured after three months for the In$_2$O$_3$/Au@C sensor, and only a slight change happened in three months, as shown in Figure 8b. It indicates that the In$_2$O$_3$/Au@C sensor also has a good long-term stability. Additionally, the resulting standard deviation is shown in Figure 8b and the number of repeat measurements is six for each concentration of H$_2$S gas.

The selectivity is very important for a gas sensor, so the In$_2$O$_3$/Au@C sensor is exposed to six other kinds of oxidizing or reducing interferential gases at room temperature with 60% RH, such as C$_3$H$_6$O, H$_2$, C$_2$H$_6$O, NH$_3$, CH$_4$, and SO$_2$, respectively. As shown in Figure 9a, the sensing sensitivity to the seven kinds of gases is 7, 3, 20, 31, 9, 41, and 228, respectively, the concentration of which is 5 ppm, including H$_2$S gas. It can be seen that the sensitivity to H$_2$S gas is highest, which is several or more than one hundred times that of other gases. As the consequence, the In$_2$O$_3$/Au@C sensor will have an excellent selective gas sensing to H$_2$S in the real environment. Further, the response curves to H$_2$S with different RH values at a concentration of 5 ppm are shown in Figure 9b at room temperature, and the sensing sensitivity is 177, 198, 228, and 206, respectively. The response curve to H$_2$S increases to the maxima with rise of the ambient RH value up to 60%, and they decreases at a higher RH. Obviously, with 60% RH the In$_2$O$_3$/Au@C sensor has the best response ability.

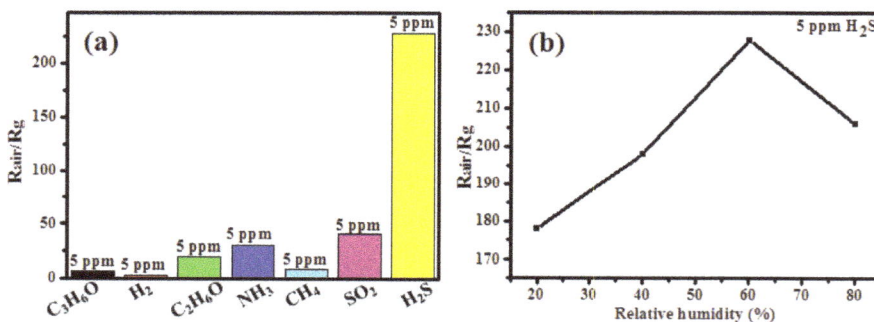

**Figure 9.** (**a**) Some potential interferential gases for the In$_2$O$_3$/Au@C sensor at room temperature with 60% RH compared to H$_2$S; and (**b**) the steady response curve to H$_2$S with a concentration of 5 ppm the at room temperature as a function of the ambient humidity from 20 to 80%.

From the above results, it is known that the Au and Au@C nanoparticles can improve the performance of In$_2$O$_3$ gas sensors. This is due to the electronic sensitization mechanism and chemical sensitization mechanism of nanoparticles, and the nanoparticles prepared by laser ablation of the metal targets in liquids have a large number of defects and chemical dangling bonds in the inner and on the surface of nanoparticles, which have higher activity [15]. On the one hand, the nanoparticles can enhance the electron density on the surface of the In$_2$O$_3$ film and adjust the resistance by the electronic sensitization mechanism. On the other hand, they have higher catalytic activity by the chemical sensitization mechanism and the spillover effect in the catalysis process [37]. They may provide more active sites for the adsorption of molecular oxygen and tracer gas. Then more electrons are provided for the redox reaction occur on the surface of In$_2$O$_3$ film, and the speed of gas sensitive reaction is accelerated. In addition, it is also observed that the In$_2$O$_3$ film modified by Au@C nanoparticles has better gas sensing properties than that modified by Au nanoparticles, indicating that the ultrathin carbon layer plays an important role on the gas sensing process. This is due to the unique electronic properties of carbon materials, such as high carrier mobility and the high sensitivity to the changes of resistance [38–40]. The ultrathin carbon layer can rapidly transform the electric carriers generated from the sensing process, and local p-n heterojunctions are created between carbon layer and In$_2$O$_3$ film, in which the carbon layer plays a role as a p-type semiconductor [37]. As a results, the performance of gas sensor is enhanced more. From the HRTEM of Au@C sample (Figure 1), it can be seen that the outer ultrathin carbon layer is about 2 nm in thickness with several carbon shells, which cannot prevent the contact of the analyte gas with the Au nanoparticles. The analyte gas can pierce through the outer carbon shells easily [41]. Finally, the more specific impact mechanisms will be studied in detail in future investigations.

## 4. Conclusions

In summary, the noble metal nanoparticles wrapped by an ultrathin carbon layer were prepared in one step based on laser ablation of the metal targets in the carbon-containing solutions. Laser irradiation of the targets not only formed the metal plasma instantly, but also the carbon atoms by breaking C-O and C-C ligands in the solution, which led to the formation of metal nanoparticles and subsequent carbon-wrapping. The thickness of the wrapped carbon layer could be tuned and controlled mainly by the carbon content in solutions, and it would reduce to ~1.3 nm by changing the proportion of toluene in the ethanol-toluene mixed solution. The $In_2O_3$ film modified by Au@C nanoparticles shows better gas sensitivity performance to $H_2S$ gas from 1 to 5 ppm at room temperature, and the ultrathin carbon layer plays an important role on the gas sensing process. This study could also be suitable for the preparation of other metal nanoparticles with wrapped ultrathin carbon layers.

**Supplementary Materials:** The following are available online at http://www.mdpi.com/2072-666X/9/6/278/s1, Figure S1: The energy spectrum of Au@C nanoparticles products by ablation of Au target in the toluene-ethanol solution with the volume ratio 9:1, Figure S2: TEM image of products induced by ablation of Au target in the water and the inset: selected area electronic diffraction of the particles, showing formation of Au nanoparticles with the size below 20 nm, Figure S3: TEM image of products induced by ablation of Ag target in the water, Figure S4: The products induced by ablation of Ag target in the toluene-ethanol solution with the volume ratio 9:1. (a): TEM image. (b): particle size distribution (data from (a)). (c): The local magnified image of (a). The inset is the selected area electronic diffraction, Figure S5: The energy spectrum of Ag@C nanoparticles products by ablation of Ag target in the toluene-ethanol solution with the volume ratio 9:1, Figure S6: The gravimetric analysis curve of the Au@C nanoparticles sample, Figure S7: The colour of the colloid solution obtained in the four solutions of (A) water; (B) pure ethanol; (C) toluene-ethanol mixed solution with the volume ratio of 1:1; and (D) toluene-ethanol mixed solution with the volume ratio of 9:1, Figure S8: The response curve (Rair/Rg) as function of test time to H2S gas with 1 ppm concentration at room temperature

**Author Contributions:** X.X. and G.D. conceived and designed the experiments; X.X. performed the experiments; L.G. analyzed the data; L.G. and X.X. wrote the paper, and they contributed equally and were co-first authors.

**Acknowledgments:** The authors acknowledge the financial supports from National Key R and D Program of China (2016YFC0201103), and financial supports from the Natural Science Foundation of China (grant no. 51471161 and 11674320), Anhui Provincial Natural Science Foundation for Distinguished Young Scholar (1408085J10), Youth Innovation Promotion Association CAS, and the Key Research Projects of the Frontier Science CAS (QYZDB-SSW-JSC017).

**Conflicts of Interest:** The authors declare no conflict of interest.

## References

1. Tan, G.Q.; Wu, F.; Yuan, Y.F.; Chen, R.J.; Zhao, T.; Yao, Y.; Qian, J.; Liu, J.R.; Ye, Y.S.; Shahbazian-Yassar, R.; et al. Freestanding three-dimensional core-shell nanoarrays for lithium-ion battery anodes. *Nat. Commun.* **2016**, *7*. [CrossRef] [PubMed]

2. Chen, X.S.; Liu, G.B.; Zheng, W.; Feng, W.; Cao, W.W.; Hu, W.P.; Hu, P.A. Vertical 2D $MoO_2$/$MoSe_2$ Core-Shell Nanosheet Arrays as High-Performance Electrocatalysts for Hydrogen Evolution Reaction. *Adv. Funct. Mater.* **2016**, *26*, 8537–8544. [CrossRef]

3. Zhu, H.; Zhang, J.F.; Yanzhang, R.P.; Du, M.L.; Wang, Q.F.; Gao, G.H.; Wu, J.D.; Wu, G.M.; Zhang, M.; Liu, B.; et al. When Cubic Cobalt Sulfide Meets Layered Molybdenum Disulfide: A Core-Shell System Toward Synergetic Electrocatalytic Water Splitting. *Adv. Mater.* **2015**, *27*, 4752–4759. [CrossRef] [PubMed]

4. Li, J.F.; Zhang, Y.J.; Ding, S.Y.; Panneerselvam, R.; Tian, Z.Q. Core-Shell Nanoparticle-Enhanced Raman Spectroscopy. *Chem. Rev.* **2017**, *117*, 5002–5069. [CrossRef] [PubMed]

5. Xu, W.G.; Xiao, J.Q.; Chen, Y.F.; Chen, Y.B.; Ling, X.; Zhang, J. Graphene-Veiled Gold Substrate for Surface-Enhanced Raman Spectroscopy. *Adv. Mater.* **2013**, *25*, 928–933. [CrossRef] [PubMed]

6.   Boies, A.M.; Lei, P.; Calder, S.; Girshick, S.L. Gas-phase production of gold-decorated silica nanoparticles. *Nanotechnology* **2011**, *22*. [CrossRef] [PubMed]

7.   Seipenbusch, M.; Binder, A. Structural Stabilization of Metal Nanoparticles by Chemical Vapor Deposition-Applied Silica Coatings. *J. Phys. Chem. C* **2009**, *113*, 20606–20610. [CrossRef]

8.   Boies, A.M.; Roberts, J.T.; Girshick, S.L.; Zhang, B.; Nakamura, T.; Mochizuki, A. $SiO_2$ coating of silver nanoparticles by photoinduced chemical vapor deposition. *Nanotechnology* **2009**, *20*. [CrossRef] [PubMed]

9.   Gao, L.J.; He, J.H. A facile dip-coating approach based on three silica sols to fabrication of broadband antireflective superhydrophobic coatings. *J. Colloid Interface Sci.* **2013**, *400*, 24–30. [CrossRef] [PubMed]

10.  Malynych, S.; Luzinov, I.; Chumanov, G. Poly(vinyl pyridine) as a universal surface modifier for immobilization of nanoparticles. *J. Phys. Chem. B* **2002**, *106*, 1280–1285. [CrossRef]

11.  Li, J.F.; Huang, Y.F.; Ding, Y.; Yang, Z.L.; Li, S.B.; Zhou, X.S.; Fan, F.R.; Zhang, W.; Zhou, Z.Y.; Wu, D.Y.; et al. Shell-isolated nanoparticle-enhanced Raman spectroscopy. *Nature* **2010**, *464*, 392–395. [CrossRef] [PubMed]

12.  Nikolov, A.S.; Balchev, I.I.; Nedyalkov, N.N.; Kostadinov, I.K.; Karashanova, D.B.; Atanasova, G.B. Influence of the laser pulse repetition rate and scanning speed on the morphology of Ag nanostructures fabricated by pulsed laser ablation of solid target in water. *Appl. Phys. A Mater. Sci. Process.* **2017**, *123*, 719. [CrossRef]

13.  Li, S.; Zhang, H.; Xu, L.L.; Chen, M. Laser-induced construction of multi-branched CuS nanodendrites with excellent surface-enhanced Raman scattering spectroscopy in repeated applications. *Opt. Express* **2017**, *25*, 16204–16213. [CrossRef] [PubMed]

14.  Bozon-Verduraz, F.; Brayner, R.; Voronov, V.V.; Kirichenko, N.A.; Simakin, A.V.; Shafeev, G.A. Production of nanoparticles by laser-induced ablation of metals in liquids. *Quantum Electron.* **2003**, *33*, 714–720. [CrossRef]

15.  Bauer, F.; Michalowski, A.; Nolte, S. Residual Heat in Ultra-Short Pulsed Laser Ablation of Metals. *J. Laser Micro Nanoeng.* **2015**, *10*, 325–328. [CrossRef]

16.  Garcia Guilen, G.; Mendivil Palma, M.I.; Krishnan, B.; Avellaneda, D.; Castillo, G.A.; Das Roy, T.K.; Shaji, S. Structure and morphologies of ZnO nanoparticles synthesized by pulsed laser ablation in liquid: Effects of temperature and energy fluence. *Mater. Chem. Phys.* **2015**, *162*, 561–570. [CrossRef]

17.  Lasemi, N.; Pacher, U.; Rentenberger, C.; Bomati-Miguel, O.; Kautek, W. Laser-Assisted Synthesis of Colloidal Ni/NiOx Core/Shell Nanoparticles in Water and Alcoholic Solvents. *Chemphyschem* **2017**, *18*, 1118–1124. [CrossRef] [PubMed]

18.  Hua, Z.Q.; Qiu, Z.L.; Li, Y.; Zeng, Y.; Wu, Y.; Tian, X.M.; Wang, M.J.; Li, E.P. A theoretical investigation of the power-law response of metal oxide semiconductor gas sensors II: Size and shape effects. *Sens. Actuator B Chem.* **2018**, *255*, 3541–3549. [CrossRef]

19.  Yin, C.Q.; Gao, L.; Zhou, F.; Duan, G.T. Facile Synthesis of Polyaniline Nanotubes Using Self-Assembly Method Based on the Hydrogen Bonding: Mechanism and Application in Gas Sensing. *Polymers* **2017**, *9*, 544. [CrossRef]

20.  Su, X.S.; Gao, L.; Zhou, F.; Duan, G.T. A substrate-independent fabrication of hollow sphere arrays via template-assisted hydrothermal approach and their application in gas sensing. *Sens. Actuator B Chem.* **2017**, *251*, 74–85. [CrossRef]

21.  Song, Z.L.; Liu, J.Y.; Liu, Q.; Yu, H.X.; Zhang, W.K.; Wang, Y.; Huang, Z.; Zang, J.F.; Liu, H. Enhanced H2S gas sensing properties based on $SnO_2$ quantum wire/reduced graphene oxide nanocomposites: Equilibrium and kinetics modeling. *Sens. Actuator B Chem.* **2017**, *249*, 632–638. [CrossRef]

22.  Zhang, S.; Song, P.; Zhang, J.; Yan, H.H.; Li, J.; Yang, Z.X.; Wang, Q. Highly sensitive detection of acetone using mesoporous $In_2O_3$ nanospher es decorated with Au nanoparticles. *Sens. Actuator B Chem.* **2017**, *242*, 983–993. [CrossRef]

23.  Park, S. Acetone gas detection using $TiO_2$ nanoparticles functionalized $In_2O_3$ nanowires for diagnosis of diabetes. *J. Alloys Compd.* **2017**, *696*, 655–662. [CrossRef]

24.  Pang, Z.Y.; Nie, Q.X.; Wei, A.F.; Yang, J.; Huang, F.L.; Wei, Q.F. Effect of $In_2O_3$ nanofiber structure on the ammonia sensing performances of $In_2O_3$/PANI composite nanofibers. *J. Mater. Sci.* **2017**, *52*, 686–695. [CrossRef]

25.  Plashnitsa, V.V.; Elumalai, P.; Kawaguchi, T.; Fujio, Y.; Miura, N. Highly Sensitive and Selective Zirconia-Based Propene Sensor using Nanostructured Gold Sensing Electrodes Fabricated from Colloidal Solutions. *J. Phys. Chem. C* **2009**, *113*, 7857–7862. [CrossRef]

26. Vargas-Garcia, R.; Romero-Serrano, A.; Angeles-Hernandez, M.; Chavez-Alcala, F.; Gomez-Yanez, C. Pt electrode-based sensor prepared by metal organic chemical vapor deposition for oxygen activity measurements in glass melts. *Sens. Mater.* **2002**, *14*, 47–56.

27. Mirzaei, A.; Janghorban, K.; Hashemi, B.; Neri, G. Metal-core@metal oxide-shell nanomaterials for gas-sensing applications: A review. *J. Nanopart. Res.* **2015**, *17*, 371. [CrossRef]

28. Bianco, G.V.; Giangregorio, M.M.; Losurdo, M.; Capezzuto, P.; Bruno, G. Supported Faceted Gold Nanoparticles with Tunable Surface Plasmon Resonance for NIR-SERS. *Adv. Funct. Mater.* **2012**, *22*, 5081–5088. [CrossRef]

29. Chen, J.Y.; Chen, Y.C. A label-free sensing method for phosphopeptides using two-layer gold nanoparticle-based localized surface plasma resonance spectroscopy. *Anal. Bioanal. Chem.* **2011**, *399*, 1173–1180. [CrossRef] [PubMed]

30. Lee, K.S.; El-Sayed, M.A. Gold and silver nanoparticles in sensing and imaging: Sensitivity of plasmon response to size, shape, and metal composition. *J. Phys. Chem. B* **2006**, *110*, 19220–19225. [CrossRef] [PubMed]

31. Norman, T.J.; Grant, C.D.; Magana, D.; Zhang, J.Z.; Liu, J.; Cao, D.L.; Bridges, F.; Van Buuren, A. Near infrared optical absorption of gold nanoparticle aggregates. *J. Phys. Chem. B* **2002**, *106*, 7005–7012. [CrossRef]

32. Boskovic, B.O.; Stolojan, V.; Khan, R.U.A.; Haq, S.; Silva, S.R.P. Large-area synthesis of carbon nanofibres at room temperature. *Nat. Mater.* **2002**, *1*, 165–168. [CrossRef] [PubMed]

33. Nemanich, R.J.; Solin, S.A. 1st-order and 2nd-order raman-scattering from finite-size crystals of graphite. *Phys. Rev. B* **1979**, *20*, 392–401. [CrossRef]

34. Huang, C.C.; Yeh, C.S.; Ho, C.J. Laser ablation synthesis of spindle-like gallium oxide hydroxide nanoparticles with the presence of cationic cetyltrimethylammonium bromide. *J. Phys. Chem. B* **2004**, *108*, 4940–4945. [CrossRef]

35. Mafune, F.; Kohno, J.; Takeda, Y.; Kondow, T.; Sawabe, H. Formation of gold nanoparticles by laser ablation in aqueous solution of surfactant. *J. Phys. Chem. B* **2001**, *105*, 5114–5120. [CrossRef]

36. Patil, P.P.; Phase, D.M.; Kulkarni, S.A.; Ghaisas, S.V.; Kulkarni, S.K.; Kanetkar, S.M.; Ogale, S.B.; Bhide, V.G. Pulsed-laser induced reactive quenching at a liquid-solid interface-aqueous oxidation of iron. *Phys. Rev. Lett.* **1987**, *58*, 238–241. [CrossRef] [PubMed]

37. Meng, F.; Zheng, H.; Chang, Y.; Zhao, Y.; Li, M.; Wang, C.; Sun, Y.; Liu, J. One-Step Synthesis of Au/SnO$_2$/RGO Nanocomposites and Their VOC Sensing Properties. *IEEE Trans. Nanotechnol.* **2018**, *17*, 212–219. [CrossRef]

38. Javey, A.; Guo, J.; Wang, Q.; Lundstrom, M.; Dai, H.J. Ballistic carbon nanotube field-effect transistors. *Nature* **2003**, *424*, 654–657. [CrossRef] [PubMed]

39. Heinze, S.; Tersoff, J.; Martel, R.; Derycke, V.; Appenzeller, J.; Avouris, P. Carbon nanotubes as Schottky barrier transistors. *Phys. Rev. Lett.* **2002**, *89*, 106801. [CrossRef] [PubMed]

40. Cong, H.; Yang, F.; Xue, C.L.; Yu, K.; Zhou, L.; Wang, N.; Cheng, B.W.; Wang, Q.M. Multilayer Graphene-GeSn Quantum Well Heterostructure SWIR Light Source. *Small* **2018**, *14*. [CrossRef] [PubMed]

41. Zhang, Y.; Li, Y.; Jiang, Y.; Li, Y.; Li, S. The synthesis of Au@C@Pt core-double shell nanocomposite and its application in enzyme-free hydrogen peroxide sensing. *Appl. Surf. Sci.* **2016**, *378*, 375–383.

*Article*

# Electrospray Deposition of ZnO Thin Films and Its Application to Gas Sensors

**Wenwang Li [1], Jinghua Lin [1], Xiang Wang [1,*], Jiaxin Jiang [2,3,4], Shumin Guo [5] and Gaofeng Zheng [2,3,4,*]**

[1] School of Mechanical and Automotive Engineering, Xiamen University of Technology, Xiamen 361024, China; xmlww@xmut.edu.cn (W.L.); ljh@stu.xmut.edu.cn (J.L.)
[2] Department of Instrumental and Electrical Engineering, Xiamen University, Xiamen 361005, China; jiangjx@xmu.edu.cn
[3] Xiamen Key Laboratory of Optoelectronic Transducer Technology, Xiamen 361005, China
[4] Fujian Key Laboratory of Universities and Colleges for Transducer Technology, Xiamen 361005, China
[5] School of Mathematical Sciences, Xiamen University, Xiamen 361005, China; shumin_guo@xmu.edu.cn
* Correspondence: wx@xmut.edu.cn (X.W.); zheng_gf@xmu.edu.cn (G.Z.); Tel.: +86-592-629-1386 (X.W.); +86-592-218-5927 (G.Z.)

Received: 24 December 2017; Accepted: 1 February 2018; Published: 2 February 2018

**Abstract:** Electrospray is a simple and cost-effective method to fabricate micro-structured thin films. This work investigates the electrospray process of ZnO patterns. The effects of experimental parameters on jet characteristics and electrosprayed patterns are studied. The length of stable jets increases with increasing applied voltage and flow rate, and decreases with increasing nozzle-to-substrate distance, while electrospray angles exhibit an opposite trend with respect to the stable jet lengths. The diameter of electrosprayed particles decreases with increasing applied voltage, and increases with flow rate. Furthermore, an alcohol gas sensor is presented. The ZnAc is calcined into ZnO, which reveals good repeatability and stability of response in target gas. The sensing response, defined as the resistance ratio of $R_0/R_g$, where $R_0$ and $R_g$ are resistance of ZnO in air and alcohol gas, increases with the concentration of alcohol vapors and electrospray deposition time.

**Keywords:** electrospray; ZnO; gas sensor; semiconductor

---

## 1. Introduction

Inkjet printing [1] has attracted attention for the fabrication of micro/nano-functional structures as it is a non-contact technology that does not require etching and exposure processing. It allows complex functional patterns to be printed precisely on flat and three-dimensional substrates. Different printing and deposition techniques have been developed. Among these, electrospray deposition [2] is of particular interest as it represents an alternative to conventional deposition and photolithographic patterning techniques for various functional structures. Electrospray is an emerging technology that utilizes electric fields to realize liquid jet ejection to fabricate micro-/nano-scale droplets, particles, and thin films [3]. The electrostatic field can cause the liquid jets to be broken into smaller particles with diameters between 10 and 500 nm by charge repulsion. Electrospray involves the use of simple equipment and has excellent material compatibility, which allows the fabrication of a diverse range of materials, including polymers [4,5], semiconductors [6], and bio-particles [7].

Zinc oxide (ZnO) [8] is a semiconductor with a wide band gap and a high electron mobility. ZnO nanostructures exhibit desirable physical and chemical properties for use in a variety of applications, including electronic devices [9], sensors [10,11] and optoelectronic devices [12]. There are a number of different methods used to fabricate ZnO films [13,14], including ultrasonic spraying, thermal evaporation, and chemical vapor deposition, etc. Among these methods, electrospray is a simple and cost-effective way to fabricate microstructured thin films. In this paper, we report the use of electrospray to deposit ZnO patterns. The effects of experimental parameters on electrospray jet characteristics and electrosprayed patterns are studied. Additionally, a gas sensor is fabricated and the sensing response is demonstrated.

## 2. Materials and Methods

Zinc acetate dihydrate (ZnAc) and glycerol are used to prepare the solution for electrospray. Briefly, the ZnAc is added to the glycerol, and the concentration of ZnAc is 4 wt %. The ZnAc/glycerol solution is stirred at room temperature for 24 h to form a homogeneous solution.

The schematic of the experimental setup is shown in Figure 1. The ZnAc/glycerol solution is loaded into a syringe equipped with a capillary steel nozzle with an inner diameter of 0.06 mm and an outer diameter of 0.11 mm. A precise syringe pump (Harvard 11 Pico Plus, Harvard Apparatus, Cambridge, MA, USA) feeds the syringe at a controllable flow rate. A high-voltage power supply (DW-P403-1AC, Tianjin Dongwen High Voltage Power Supply Plant, Tianjin, China) with the anode connected to the nozzle and the cathode connected to a grounded glass substrate is used to generate the electric field for electrospray. The morphology of electrosprayed patterns is examined using an optical microscope.

The ZnAc/glycerol solution is served as a precursor. To generate the ZnO thin film, the electrosprayed ZnAc pattern is placed on a heating plate (C-MAG HS7, IKA®-Werke GmbH & Co. KGStaufen, Breisgau, Germany) to conduct a calcination process. This calcination process is conducted at a temperature of 773 K in air. The heating rate is set at 10 K/min and the heating time is 1 h. After the calcination, the ZnAc/glycerol is decomposed and oxidized to obtain ZnO.

The electrospray process is observed and captured by a CCD camera (UI-1250SE, IDS Imaging Development Systems GmbH, Obersulm, Germany) equipped with a zoom lens. The electrosprayed patterns are observed by an optical microscope (FS-70, Mitutoyo Co., Kawasaki, Japan). The stable jet length, electrospray angle and particle diameter are measured with the assistance of an image processing and analysis software (ImageJ, National Institutes of Health, Bethesda, MD, USA).

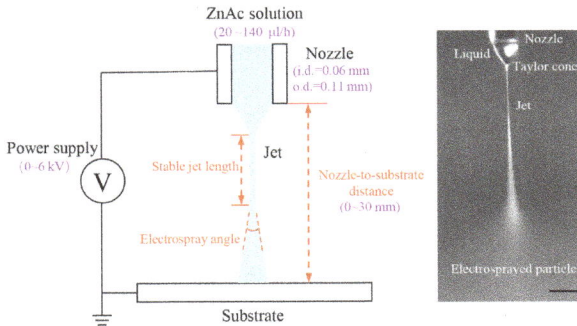

**Figure 1.** Schematic of the experimental setup. The optical image inset illustrates the electrospray process of ZnAc solution; scale bar: 0.1 mm.

## 3. Results and Discussion

In the typical electrospray apparatus, a high voltage is applied to a fluid supplied though a capillary nozzle. The liquid drop on the nozzle tip subjects an electrostatic force on the interface as a result of electric field generated between the nozzle and collector. This electrostatic force opposes the surface tension force and deforms the liquid into a conical shape, called Taylor cone [15]. If the electric field applied is large enough, the electric field force will overcome the surface tension force, results a liquid jet ejected through the apex of the Taylor cone to the collector. The liquid jet travels a straight path and then breaks into small charged liquid droplets/particles which are radially dispersed due to the Coulomb repulsion [16].

Firstly, the influence of experimental parameters on the electrospray process is investigated. Figure 2 illustrates the effect of the applied voltage on the length of the stable jets and on the electrospray angles at a nozzle-to-substrate distance of 20 mm and a flow rate of 20 µL/h. The electrospray process maintains a stable cone-jet ejection when the applied voltage is above 3 kV. The length of stable jets increases with increasing applied voltage and the electrospray angle decreases when the applied voltage is enlarged. The stable jet lengths at applied voltages of 4, 4.5, 5, and 5.5 kV are 0.148, 0.153, 0.167, and 0.180 mm, and the corresponding electrospray angles are 79°, 78°, 55°, and 37°, respectively. Theoretically, the axial electric field intensity increases with increasing applied voltage. Under such conditions the liquid experiences larger electric field forces. In this case, both axial acceleration, as well as axial velocity increase, resulting in longer stable jets and smaller electrospray angles. Although the radial electric field also increases with the applied voltage, the axial field dominates under such experimental conditions.

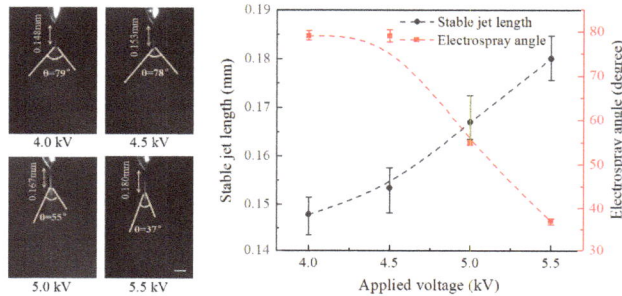

**Figure 2.** Effect of applied voltage on the stable jet length and electrospray angle. The nozzle-to-substrate distance is 20 mm and the flow rate is 20 µL/h. Scale bar: 0.1 mm.

Figure 3 shows the effect of flow rate on the stable jet lengths and on the electrospray angles at a fixed nozzle-to-substrate distance of 20 mm and an applied voltage of 6 kV. It can be seen that the stable jet lengths increase with increasing flow rate, while the electrospray angles decrease with the flow rate. The stable jet lengths at flow rates of 20, 60, 100, and 140 µL/h are 0.110, 0.120, 0.134, and 0.161 mm, and the corresponding electrospray angles are 97°, 73°, 60°, and 51°, respectively. Increasing flow rates accelerate the liquid ejection, thus increasing the stable jet lengths and decreasing the electrospray angles.

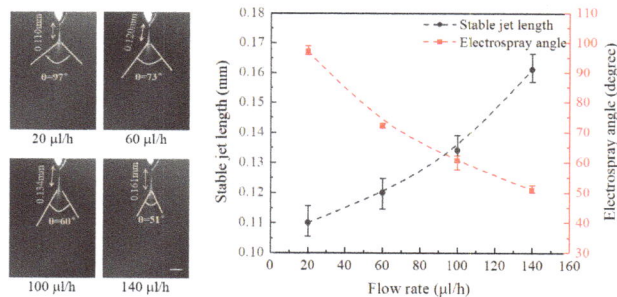

**Figure 3.** Effect of flow rate on the stable jet length and electrospray angle. The nozzle-to-substrate distance is 20 mm and the applied voltage is 6 kV. Scale bar: 0.1 mm.

Figure 4 shows the effect of nozzle-to-substrate distance on the stable jet length and the electrospray angle at an applied voltage of 6 kV and a flow rate of 20 μL/h. It reveals that the length of stable jets decreases with increasing nozzle-to-substrate distance, while the electrospray angle increases with the nozzle-to-substrate distance. Stable jet lengths at nozzle-to-substrate distances of 20, 25, and 30 mm are 0.245, 0.208, and 0.132 mm, and the corresponding electrospray angles are 60°, 74°, and 79°, respectively. Increasing the nozzle-to-substrate distances reduces the electric field intensity, the force acting on the liquid is decreased such that the nozzle-to-substrate distance is shortened, and the electrospray angle is enlarged.

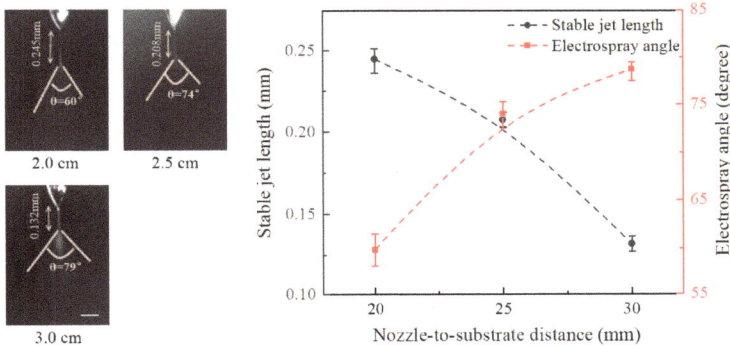

**Figure 4.** Effect of nozzle-to-substrate distance on the stable jet length and electrospray angle. The applied voltage is 6 kV and the flow rate is 20 μL/h. Scale bar: 0.1 mm.

The experimental parameters also impact the morphology of the deposited patterns. Figure 5 shows the morphology of deposited patterns under various applied voltages. The nozzle-to-substrate distance and the flow rate are 10 mm and 20 μL/h, respectively. The average diameter of the electrosprayed particles decreases with increasing applied voltage. The average diameter is 4.51, 4.07, 3.27, and 2.88 μm when the applied voltage is 3.5, 4.5, 5.5, and 6.5 kV, respectively. Actually, increasing the applied voltages increases the electric field intensity and the charge density on the solution surface. Increasing the surface charge density increases the Coulomb repulsion within the ejected liquid, which contributes to the atomization of the charged jets.

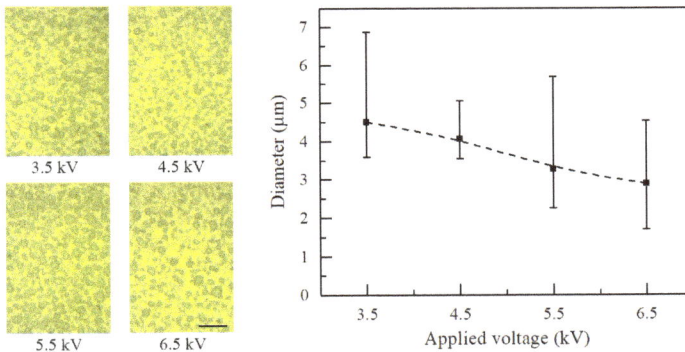

**Figure 5.** Effect of applied voltage on the diameter of the electrosprayed particles. The nozzle-to-substrate distance and the flow rate are 10 mm and 20 μL/h, respectively. The optical photographs inset are the electrosprayed particles. Scale bar: 20 μm.

Figure 6 shows the morphology of the deposited patterns under various flow rates at a nozzle-to-substrate distance of 10 mm and an applied voltage of 2.5 kV, indicating an increase in the average diameter of electrosprayed particles with increasing flow rate. The average diameter is 7.42, 12.09, 13.10, 13.78, and 14.24 μm when the flow rate is 20, 40, 60, 80, and 100 μL/h, respectively. As mentioned above, large flow rates favor fast liquid ejection. The charged jet do not have enough time to atomize under the short nozzle-to-substrate distance, resulting in larger electrosprayed particles.

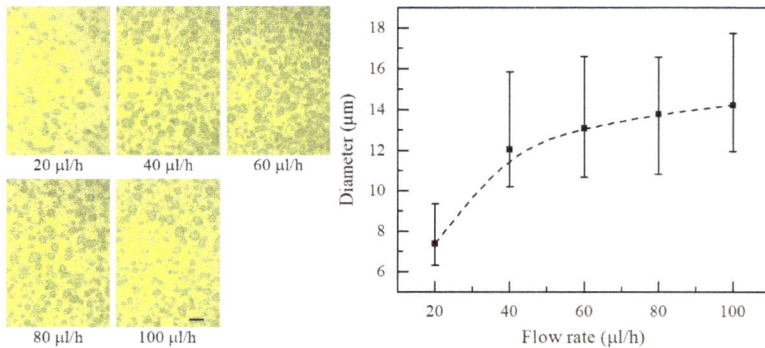

**Figure 6.** Effect of the flow rate on the diameter of electrosprayed particles. The nozzle-to-substrate distance and the applied voltage are 10 mm and 2.5 kV, respectively. The optical photographs inset are the electrosprayed particles. Scale bar: 20 μm.

To demonstrate the feasibility of electrospray processes in the field of sensor manufacturing, an alcohol gas sensor using ZnO patterns as a sensitive layer is fabricated on a silicon substrate. This substrate is pre-processed to form two electrodes with a gap of 1 mm. The ZnAc solution is electrosprayed over the gap between the two Au electrodes, and then calcined to form ZnO structures. The applied voltage, flow rate, and nozzle-to-substrate distance are 1.5 kV, 50 μL/h, and 0.5 mm, respectively. A diagrammatic sketch of the prepared gas sensor and the measuring system is shown in Figure 7. The two electrodes of the alcohol gas sensor are connected to the anode and cathode of the DC voltage supply (GPC3060D, Gwinstek, New Taipei City, Taiwan), and the current passing through the sensor is measured by a digital multimeter (34410A, Agilent Technologies, Santa Clara, CA, USA). The changes in the ZnO resistance in alcohol vapor and in air are evaluated to calculate the sensing response. The sensing response is defined as $R_0/R_g$, where $R_0$ and $R_g$ are the resistances of ZnO in air and in alcohol vapor, respectively.

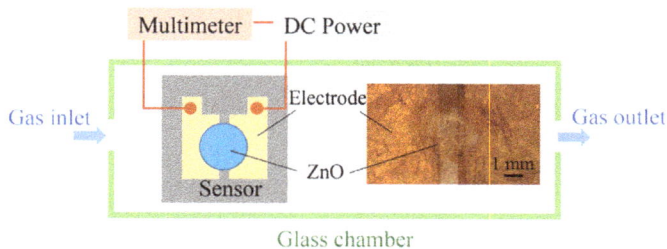

**Figure 7.** Diagrammatic sketch of the prepared gas sensor and the measuring system. The gap between the two electrodes is about 1 mm. The diameter of ZnO deposition area is in the range of 1.5–2.5 mm.

When the ZnO is exposed to alcohol vapor, the alcohol molecules adsorb on the ZnO surface, resulting in a decreased resistance. Figure 8 shows the sensing response of ZnO towards repeated exposures to 150 ppm alcohol vapor. The deposition time for electrospray is 15 min. The resistance of the sensitive material in air and alcohol are about $1.55 \times 10^8$ Ω and $0.43 \times 10^7$ Ω, respectively. The sensing response $R_0/R_g$ is about 36. Moreover, it reveals a good repeatability and stability of response in the target gas. Figure 9 shows the sensing response of ZnO thin films formed at increasingly larger electrospray times as a function of alcohol concentration. Higher vapor concentrations cause more molecules to absorb on the ZnO surface, thus resulting in a resistance decrease. Thus, $R_0/R_g$ increases with increasing alcohol concentration. On the other hand, $R_0/R_g$ increases with increasing electrospray deposition time. Longer ejection periods lead to larger amounts of sensor elements, as the thickness of the ZnO deposition increases with increasing electrospray time. The diameter of ZnO deposition area is in the range about 1.5–2.5 mm. The average thicknesses for various deposition times of 5, 10, 15, and 20 min are estimated to be 100, 130, 180, and 200 nm, respectively.

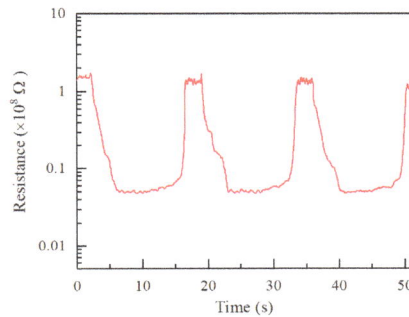

**Figure 8.** Response of the gas sensor towards repeated exposures to the alcohol vapor. The deposition time for ZnO is 15 min. The concentration of alcohol is 150 ppm.

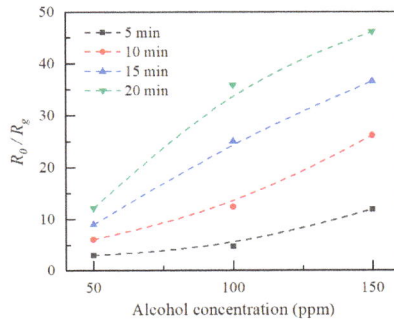

**Figure 9.** Sensing response of ZnO thin films formed at increasingly longer electrospray times and exposed to increasingly large alcohol concentrations.

## 4. Conclusions

This study investigates the effect of experimental parameters on the electrospray process of ZnAc. We have shown that the stable jet length increases with increasing applied voltage and flow rate, while it decreases with increasing nozzle-to-nozzle substrate distance. The corresponding electrospray angles show an opposite in contrast to the stable jet lengths. The diameter of electrosprayed particles decreases with increasing applied voltage, and increases with the flow rate. Furthermore, an alcohol gas sensor has been demonstrated by electrospray and calcining processes. The ZnAc is calcined into

*Micromachines* **2018**, *9*, 66

ZnO, which reveals good repeatability and stability of response in the target gas. The sensing response, defined as the resistance ratio of $R_0/R_g$, increases with the concentration of alcohol vapors and the deposition time of electrospray.

**Acknowledgments:** This work is supported by the National Natural Science Foundation of China (51575464, U1505243), the Education and Scientific Research Projects for Middle-aged and Young Teachers of Fujian Province of China (JAT160359), Science and Technology Planning Project of Fujian Province (2017H0037), and the Xiamen Municipal Science and Technology Projects (3502Z20163005).

**Author Contributions:** W.L., X.W., and G.Z. conceived the idea. W.L., J.L., J.J., and S.G. provided the data. W.L., X.W., and G.Z. analyzed the data. All authors contributed to the writing and revisions.

**Conflicts of Interest:** The authors declare no conflict of interest.

## References

1. Yin, Z.; Huang, Y.; Bu, N.; Wang, X.; Xiong, Y. Inkjet printing for flexible electronics: Materials, processes and equipments. *Chin. Sci. Bull.* **2010**, *55*, 3383–3407. [CrossRef]
2. Jaworek, A.; Sobczyk, A.T. Electrospraying route to nanotechnology: An overview. *J. Electrostat.* **2008**, *66*, 197–219. [CrossRef]
3. Li, W.; Zheng, G.; Xu, L.; Wang, X. Fabrication of micro-patterns via near-field electrospray. *AIP Adv.* **2016**, *6*, 115002. [CrossRef]
4. Park, C.H.; Lee, J. Electrosprayed Polymer Particles: Effect of the Solvent Properties. *J. Appl. Polym. Sci.* **2009**, *114*, 430–437. [CrossRef]
5. Cai, X.; Lei, T.; Sun, D.; Lin, L. A critical analysis of the α, β and γ phases in poly(vinylidene fluoride) using FTIR. *RSC Adv.* **2017**, *7*, 15382–15389. [CrossRef]
6. Ju, J.; Yamagata, Y.; Higuchi, T. Thin-film fabrication method for organic light-emitting diodes using electrospray deposition. *Adv. Mater.* **2009**, *21*, 4343–4347. [CrossRef] [PubMed]
7. Sahoo, S.; Lee, W.C.; Goh, J.C.H.; Toh, S.L. Bio-Electrospraying: A Potentially Safe Technique for Delivering Progenitor Cells. *Biotechnol. Bioeng.* **2010**, *106*, 690–698. [CrossRef] [PubMed]
8. Schmidtmende, L.; Macmanusdriscoll, J.L. ZnO—Nanostructures, defects, and devices. *Mater. Today* **2007**, *10*, 40–48. [CrossRef]
9. Wang, X.; Zheng, G.; He, G.; Wei, J.; Liu, H.; Lin, Y.; Zheng, J.; Sun, D. Electrohydrodynamic direct-writing ZnO nanofibers for device applications. *Mater. Lett.* **2013**, *109*, 58–61. [CrossRef]
10. Das, S.N.; Kar, J.P.; Choi, J.; Lee, T.I.; Moon, K.; Myoung, J. Fabrication and Characterization of ZnO Single Nanowire-Based Hydrogen Sensor. *J. Phys. Chem. C* **2010**, *114*, 1689–1693. [CrossRef]
11. Dhahri, R.; Leonardi, S.G.; Hjiri, M.; Mir, L.E.; Bonavita, A.; Donato, N.; Iannazzo, D.; Neri, G. Enhanced performance of novel calcium/aluminum co-doped zinc oxide for co2 sensors. *Sensor. Actuators B Chem.* **2017**, *239*, 36–44. [CrossRef]
12. Olson, D.C.; Shaheen, S.E.; Collins, R.T.; Ginley, D.S. The effect of atmosphere and ZnO morphology on the performance of hybrid poly(3-hexylthiophene)/ZnO nanofiber photovoltaic devices. *J. Phys. Chem. C* **2007**, *111*, 16670–16678. [CrossRef]
13. Kumar, R.; Kumar, G.; Al-Dossary, O.; Umar, A. ZnO nanostructured thin films: Depositions, properties and applications—A review. *Mater. Express* **2015**, *5*, 3–23. [CrossRef]
14. Angelini, E.; Grassini, S.; Rosalbino, F.; Fracassi, F.; Laera, S.; Palumbo, F. PECVD coatings: Analysis of the interface with the metallic substrate. *Surf. Interface Anal.* **2010**, *38*, 248–251. [CrossRef]
15. De La Mora, J.F. The fluid dynamics of Taylor cones. *Annu. Rev. Fluid Mech.* **2007**, *39*, 217–243. [CrossRef]
16. Rohner, T.C.; Lion, N.; Girault, H.H. Electrochemical and theoretical aspects of electrospray ionisation. *Phys. Chem. Chem. Phys.* **2004**, *6*, 3056–3068. [CrossRef]

MDPI

St. Alban-Anlage 66

4052 Basel

Switzerland

Tel. +41 61 683 77 34

Fax +41 61 302 89 18

www.mdpi.com

*Micromachines* Editorial Office

E-mail: micromachines@mdpi.com

www.mdpi.com/journal/micromachines

www.ingramcontent.com/pod-product-compliance
Lightning Source LLC
Chambersburg PA
CBHW051917210326
41597CB00033B/6171